# The Global Warming Combat Manual

# The Global Warming Combat Manual

## Solutions for a Sustainable World

BRUCE E. JOHANSEN

*Foreword by James E. Hansen*

Westport, Connecticut
London

333.79
Joh

**Library of Congress Cataloging-in-Publication Data**

Johansen, Bruce E. (Bruce Elliott), 1950–
    The global warming combat manual : solutions for a sustainable world / Bruce E. Johansen ; foreword by James E. Hansen.
        p. cm.
    Includes bibliographical references and index.
    ISBN: 978-0-313-35286-7 (alk. paper)
    1. Environmental protection—Citizen participation. 2. Global warming—Prevention—Citizen participation. I. Title.
TD171.7.J64 2008
333.79—dc22        2008008987

British Library Cataloguing in Publication Data is available.

Library of Congress Catalog Card Number: 2008008987
ISBN: 978-0-313-35286-7

First published in 2008

Praeger Publishers, 88 Post Road West, Westport, CT 06881
An imprint of Greenwood Publishing Group, Inc.
www.praeger.com

Printed in the United States of America

The paper used in this book complies with the Permanent Paper Standard issued by the National Information Standards Organization (Z39.48-1984).

10  9  8  7  6  5  4  3  2  1

# Contents

# Foreword

Bruce Johansen is the most prolific and down-to-earth writer among all the authors that I know who have chosen to focus on the topic of global warming. He has been at it for a long time, even back when global warming was only of interest to the sandals and granola crowd. Now the topic is in the news every day, and even politicians, except for the most well-oiled, agree that we must deal with the problem.

I am glad to see that in his new book, *The Global Warming Combat Manual*, Bruce has turned his attention to the actions that we all can take to help preserve our climate and allow all the creatures of nature to continue to thrive on our planet. There are many things that we can do in our daily lives to reduce emissions of the greenhouse gases that cause global warming, but perhaps the most important thing that we can do is make clear to government officials that we want policies that favor our children, grandchildren, and all the species that live with us on Earth, not the short-term profits of fossil fuel special interests.

Coincidentally, when I received Bruce's book, and was asked to write this foreword, I was writing a letter to the governor of a Midwestern state. The letter defines why a moratorium on construction of coal-fired power plants is the single most critical action needed to solve the global warming problem. I hope that you will read and think about the matters discussed in this letter. It is only pressure from the public that can overcome the great influence of the special interests.

Dear Governor,

I write to you, a fellow grandparent, about a matter that concerns our children and grandchildren. The matter also has near-term ramifications.

I grew up and was educated in the Midwest and, for reasons to become apparent, I note that my political inclinations are those of a moderate conservative. I am registered as an Independent. Such political stripes may be a product of having worked my way through school (with generous help of a state,

Iowa, that forgave tuition) and lived in New York City when it was run by L-word politicians (I was held up twice at knife-point and, like other citizens, at times I needed to skirt piles of uncollected garbage and snow drifts—though, in fairness, failure to deliver services is not restricted to L-word politicians).

Over the past 40 years I have acquired knowledge about causes and consequences of climate change. Climate change, natural or human-induced, recognizes no political boundaries. Policies addressing climate change should be bipartisan. This will not be easy, as the subject has become politicized. But, with leadership, it is possible.

Just over a year ago I participated in a meeting of scientists and evangelical leaders. There was a unanimous consensus in favor of stewardship, the need to preserve creation, the planet on which civilization developed over recent millennia. Of course, not all evangelicals were present; it has been suggested that some believe it is permissible to trash the planet and nonhuman residents, anticipating that humans will be rescued at the next coming of Christ.

Yet a majority of people, especially young people, feel strongly that we have an obligation to preserve the wonders of Earth and allow the other creatures on the planet to flourish. They seek and will support political leaders who put aside partisanship for the common good.

Climate change, to the nonscientist, does not seem to be an urgent matter. Global warming in the past century is about 1°F, about 2°F over land areas. This is much smaller than weather fluctuations or even seasonal mean temperature anomalies. Yet there is at least that much additional warming "in the pipeline," due to human-made gases already in the air. Full warming is slow to appear because of the great inertia of the climate system, especially the ocean.

As a result, global climate is nearing critical tipping points that could cause: loss of Arctic sea ice with detrimental effects on wildlife and indigenous people, Antarctic and Greenland ice sheet disintegration with sea level rise possibly accelerating out of control, reduced freshwater supplies for hundreds of millions of people, and a more intense hydrologic cycle with stronger droughts and forest fires, but also heavier rains and floods, and stronger storms driven by latent heat, including tropical storms, tornados, and thunderstorms.

Effects of global warming are already becoming apparent in the United States, as well as in the Earth's polar regions. Expansion of subtropical climate, adding to increased demands for fresh water, threatens to make water bodies such as Lake Mead and Lake Powell run dry. Yet it is possible to halt and reverse these effects. America has the potential to lead the world in finding solutions to global warming. In doing so, it will create many good high-pay jobs for Americans.

Good new jobs will be associated with renewable energies such as solar power and wind power. Our desert areas in the Southwest have the potential to provide a great amount of energy as the price of carbon emissions rises inevitably and the cost of harvesting solar energy declines. Wind energy is abundant in many states. Our single largest potential source of "new" energy is improved energy efficiency. Production of products with higher energy efficiency will create many good jobs.

Our farm areas can be rejuvenated, as they help us solve the climate problem. In coming years farms can not only produce renewable energies, but help draw down the amount of carbon dioxide in the air. The most effective natural way to remove atmospheric $CO_2$ is to increase storage of carbon in the soil with appropriate agricultural practices. Just as polluters must be made to pay a price for carbon emissions, so farmers should be paid an equal price for carbon stored in the soil.

Corn-based ethanol does little if any good for the planet. But perennial cellulosic crops such as natural grasses have much great potential. Appropriate land management can increase carbon storage and reduce emission of greenhouse gases such as nitrous oxide and methane. Farms must play an important role as the world addresses climate change in an effective way.

I have written this long preamble, with a frank discussion of my own political inclinations, because I believe that the climate change issue has been inappropriately politicized. Conservatives have as much interest as anyone in preserving the environment. It is important that we have leaders who will reach across the aisle and work together to find solutions.

I hope that you will exert your leadership to mold bi-partisan cooperation needed to develop clean energies, a stronger economy, and a healthy planet, with a stabilized atmospheric composition that avoids dangerous human-made climate change. Your children and grandchildren will greatly appreciate your foresight and leadership.

There are scientific facts about fossil fuels that need to be communicated to other leaders and the public, as laid out in the attachment to this letter in the appendix. The most important science that must be communicated concerns the lifetime of fossil fuel $CO_2$ in the air, and the implication of that fact for use of specific fossil fuels. Specifically, because:

(1) a large fraction of $CO_2$ emitted in burning fossil fuels stays in the air more than 1000 years,

(2) oil reserves will be exploited sooner than that, indeed this century, even if we rapidly improve energy efficiencies, and

(3) because it is impractical to capture $CO_2$ emissions from vehicles, it becomes apparent that oil use alone will take the world to well over 400 ppm (parts per million) $CO_2$. $CO_2$ has already increased from its pre-industrial amount of 280 ppm to 385 ppm today.

Observations of ongoing climate change make clear that such $CO_2$ amounts pose grave risks of setting in motion climate change that begins to run out of our control, including nonlinear collapse of ice sheets and collapse of ecosystems that contain innumerable irreplaceable species. In view of the magnitude of the oil, gas and coal carbon reservoirs, with coal by far the largest, it becomes apparent that the one way, the only practical way, to keep $CO_2$ in the neighborhood of 400 ppm is to phase out coal use except and until the technology is available to capture and safely sequester $CO_2$ emitted in coal burning.

The attached discussion makes clear that a prompt moratorium on construction of power plants that do not capture $CO_2$ is the most important action required to solve the climate change problem. Together with improved agricultural and forestry practices that sequester carbon in the soil and biosphere, it will then be possible to restore a healthy climate, for the benefit of all people and species.

I recognize the doubts and resistance that you would face in supporting such a bold step, but the situation is urgent. A coal moratorium could be the tipping point, a first step that helps lead our country onto a path that is so clearly essential from a scientific perspective. Conversely, each new power plant without carbon capture will extinguish irreplaceable species and impoverish our children and future generations.

In trying to communicate the scientific story, and the needed actions that it implies, a basic conflict has become starkly apparent to me—the conflict between fossil fuel special interests and related industry, on the one hand, and the young people, creatures, and nature that will suffer the consequence of accelerating climate change. Young people may seem over-matched in conflict against powerful special interests. They are eagerly hoping for your assistance.

With all respect and best wishes,
James E. Hansen
Director, NASA Goddard Institute for Space Studies*

---

*Institutional affiliation is for identification purposes only. Statements relating to policy are personal opinions.

# Preface

Soon after I first met him in 2006, Tim Rinne of Nebraskans for Peace presented me with a haunting thought. During the 1960s, when the major threat to world peace was the looming threat of nuclear war, Rinne realized that such a conflict could not be initiated without affirmative human intervention, a break in the usual humdrum of life. Someone had to press the legendary red button. Global warming is so much more dangerous, he said, by virtue of its insidious nature. All we have to do to destroy our home planet's environment as we know it via global warming is to carry on business as usual, one pump of the gas pedal (or one flick of a light switch) at a time.

Bill McKibben commented in the *Washington Post*:

> Consider the news from the real world, the one where change is measured with satellites and thermometers, not focus groups.... Shaken scientists see every prediction about the future surpassed by events. As Martin Parry, co-chair of the Intergovernmental Panel on Climate Change, told reporters this month, "We are all used to talking about these impacts coming in the life-times of our children and grandchildren. Now we know that it's us." (McKibben 2007)

After every conversation about this issue, people arrive at the same destination: What do we change, how, and why? People who have been reading my books on global warming have been raising a chorus: You've scared us to death—now what do we do? This is a book about what you can do, as well as how and why you should do it. Rajendra Pachauri, chairman of the Intergovernmental Panel on Climate Change, said, "We need a new ethic by which every human being realizes the importance of the challenge we are facing and starts to take action through changes in lifestyle and attitude" (Environment News Service, 2007t).

Global warming will not have a single neat and tidy solution. The path will have many roads, a mosaic, a vast banquet involving billions of individual choices, some of which already are emerging, others that are still over the horizon of imagination. The energy revolution is already under way, in a quiet fashion: Americans in 2005 used 47 percent less energy per unit (in inflation-adjusted dollars) of economic output than they did thirty years earlier (Steinman 2007, 271). Wind and solar energy are expanding at breakneck speed as I write.

Vaclav Havel, former president of the Czech Republic, commented that "maybe we should start considering our sojourn on earth as a loan."

> There can be no doubt that for the past hundred years at least, Europe and the United States have been running up a debt, and now other parts of the world are following their example. Nature is issuing warnings that we must not only stop the debt from growing but start to pay it back. There is little point in asking whether we have borrowed too much or what would happen if we postponed the repayments. Anyone with a mortgage or a bank loan can easily imagine the answer. (Havel 2007)

Here I describe in detail practical measures readers can take in their daily lives to reduce their carbon footprints, while at the same time illustrating how these individual, daily choices link with larger legal, political, economic, and technological changes that are being pursued to deal with global warming. Dovetailing the personal with the technological and public-policy dimensions, this book lays out in detail a battery of solutions for global warming. These range from the humdrum and easy (keeping your tires properly inflated) through the necessary, long-range, and difficult (retooling the ways we transport, house, and feed ourselves for maximum energy efficiency and minimum greenhouse impact). Our journey includes the possible (switching over a large fraction of our carbon-based energy sector to alternative sectors based on biofuel, wind, solar, and geothermal power), the visionary (creating a bacterium that will consume $CO_2$), and the improbable (deploying giant reflecting mirrors out in space), all the way to the weird and dangerous (pumping sulfur aerosols into the stratosphere).

## THE URGENCY OF THE PROBLEM

The necessity of a paradigm change in how we acquire and use energy becomes more evident by the day. Scientific research has been indicating that climate models have been too conservative. The climate is changing more quickly than was previously anticipated. Diplomacy is slow, and people react only after observing problems. To deal with global warming, we not only have to react to a slow-motion crisis but, because feedbacks guarantee that we will not feel the full effect until decades after greenhouse-gas emissions provoke it, we must anticipate the outcome. Our societal nervous system does not perform well under such conditions. In the meantime, carbon dioxide levels in the atmosphere have been racing upward at the fastest pace since they have been measured (beginning in 1959), fed mainly by rapid industrialization in China and India (Canadell et al. 2007).

Evidence of global warming's accelerating pace is all around us. For example, the Arctic sea ice extent in September (at the end of the melt season, when ice coverage is at its annual minimum) has declined sharply over the past several decades. During the summer of 2007, the Arctic ice cap lost a stunning 24 percent of its mass in *one* year. Within a decade or two, we may see open water in summer in the Arctic on such a routine basis that it may scarcely merit mention. At the same time, the rapidly warming Arctic's permafrost will be injecting more and more carbon dioxide and methane into the atmosphere, as everyone comes to realize that humankind's contribution has become a trigger on a very large gun. At that time—the "tipping point"—warming may accelerate beyond any human ability to intervene.

As George W. Bush's White House bantered about "sound science" (and tried to muzzle scientists and censor reports) yellow jacket wasps were sighted on northern Baffin Island during the summer of 2004. By the end of the twenty-first century, if "business as usual" fossil-fuel consumption is not curbed substantially, the atmosphere's carbon dioxide level will reach 800 to 1,000 parts per million. The last time the level was that high, 65 million years ago, was during the notably toasty days of the dinosaurs.

Increasing evidence also indicates that rising temperatures are changing the hydrological cycle, helping to cause intensifying chances of precipitation extremes of drought and deluge. Western Europe has experienced flooding rains while the western interior of North America and Australia have been suffering drought that may be the worst since the one that ruined the civilization of the Anasazi a thousand years ago. Intensity of storms often increases with warmth; in the midst of drought during the summer of 2002, for example, sections of Nebraska experienced cloudbursts that eroded soil and washed out an interstate highway. Hours later, the drought returned.

## THE BILL WILL NOT ARRIVE UNTIL THE PRESENT GENERATION HAS LEFT THE TABLE

The reality of global warming is more complex than sweat on one's brow during a summer afternoon. Global warming is a sneaky, slow-motion, insidious threat. We as citizens of the world react to what we see and feel, and our diplomatic, legal, political, and economic actions follow those perceptions. Temperatures rise much more rapidly in winter than in summer, and at night than during the day. In addition—and this is a very important point—today's greenhouse-gas emissions do not become tomorrow's heat. Through feedback loops in the air and oceans, it will be decades, perhaps a century, "before the oceans and the atmosphere fully redistribute the absorbed energy and the currently 'committed' temperature rise is actually 'realized'" (National Academy of Sciences 1991, 93).

The *equilibrium temperature* is the level at which the system will come to rest once the full effects of greenhouse warming become evident. The "bill" for emissions in 2007 probably will fall due about 2050. That's the feedback for the air. For the oceans, it is much longer, several centuries. Sea-level rise today has barely caught up with the first few decades following the advent of the industrial revolution. Assuming present-day business as usual, within the lives of our great-grandchildren (about a century) enough sea-level rise will be "in the pipeline"—about twenty-five meters—to wreak havoc with the lives of a billion people in coastal cities around the world.

Thus the need for an emphatic emphasis on solutions now. James Hansen, who heads the Goddard Institute for Space Studies, tells us that we have a decade or two to get extremely serious about reducing greenhouse-gas emissions before feedbacks accelerate from such things as albedo (reflectivity) changes and melting permafrost in the Arctic.

## FACING OUR FEARS

In the future, students of history may remark at the nature of the fears that stalled responses to climate change early in the twenty-first century. For crucial years, those who denied the seriousness of global warming kept change at bay, it may be remarked, by appealing to most people's fear of change that might erode their comfort and employment security, all of which they associated with massive burning of fossil fuels.

Technological change always generates fear of unemployment. Paradoxically, such changes also always generate economic activity. A change in our basic energy paradigm during the twenty-first century will not cause the ruination of our economic base, as some climate-change contrarians believe, any more than the coming of the railroads in the nineteenth century ruined an economy in which the horse was the major land-based vehicle of transportation. The advent of the automobile early in the twentieth century propelled economic growth. So has the transformation of information technology using computers during recent decades. The same developments also put out of work blacksmiths, keepers of hand-drawn accounting ledgers, and anyone who repaired manual typewriters.

We are overdue for an energy system paradigm shift. Limited supplies of oil and their location in the volatile Middle East, along with accelerating climate change, has created a political and economic environment that should *welcome* such change. According to an editorial in *Business Week*:

> A national policy that cuts fossil-fuel consumption converges with a geopolitical policy of reducing energy dependence on Middle East oil. Reducing carbon dioxide emissions is no longer just a "green" thing. It makes business and foreign policy sense, as well.... In the end, the only real solution may be new energy technologies. There has been little innovation in energy since the internal combustion engine was invented in the 1860s and Thomas Edison built his first commercial electric generating plant in 1882. (2004, 108)

As of 2007, the federal government of the United States (which, as a nation, produces more than one-fifth of the world's greenhouse gases) was just beginning to warm to the next worldwide energy revolution. The United States still was being led (if that is the word) by people whose minds still were set to the clock of the early twentieth-century fossil-fuel boom.

The Bush administration not only refused to endorse the Kyoto Protocol but also failed to take seriously the coming revolution in the technology of energy production and use. In a century, George Bush's bust may sit in a Greenhouse Gas Museum, not far from a model of an antique internal combustion engine. A plaque may mention his family's intimate ties to the oil industry as a factor in his refusal to think outside that particular "box."

## NECESSITY WILL COMPEL INVENTION

Before the end of the present century, the urgency of global warming will become manifest to everyone. Solutions to our fossil-fuel dilemma— solar, wind, hydrogen, and other forms of power—will evolve during this century. Within our century, necessity will *compel* invention. Other technologies may develop that have not, as yet, even broached the realm of present-day science fiction, any more than digitized computers had in the days of the Wright Brothers a hundred years ago. We will take this journey because the changing climate, along with our own innate curiosity and creativity, will compel a changing energy paradigm.

Such change will not take place all at once. A paradigm change in basic energy technology may require the better part of a century, or longer. Several technologies will evolve together. Oil-based fuels will continue to be used for purposes that require them. A wide variety of solutions are being pursued around the world. Already, several U.S. states are taking actions to limit carbon dioxide ($CO_2$) emissions despite a lack of support from the federal government. Building code changes have been enacted. Wind power is exploding across the United States. In Texas. some former oil fields now host wind turbines. Wind turbines and solar cells are becoming more efficient and competitive. Improvements in farming technology are reducing emissions. Deep-sea sequestration of $CO_2$ is proceeding in experimental form, but with concerns about this technology's effects on ocean biota. Tokyo, where an urban heat island has intensified the effects of general warming, has proposed a gigantic ocean-water cooling grid. Britain and other countries are considering carbon taxes.

The coming energy revolution will engender economic growth and become an engine of wealth creation for those who realize its opportunities. Denmark, for example, is making every family a shareowner in a burgeoning wind-power industry. The United Kingdom is making plans to reduce its greenhouse-gas emissions 50 percent in 50 years. The British program begins to address the position of the Intergovernmental Panel on Climate Change (IPCC) that emissions will have to fall by 80 percent or more by century's end to avoid significant warming of the lower atmosphere due to human activities. The Kyoto Protocol, with its reductions of 5 to 15 percent (depending on the country) is barely earnest money compared to the required paradigm change that will fundamentally transform the system by which most of the world's people obtain and use energy.

Solutions will combine scientific innovation and political change. We will end this century with a new energy system, one that acknowledges nature and works with its needs and cycles. Economic development will become congruent with the requirements of sustaining nature. Coming generations will be able to mitigate the effects of greenhouse gases without the increase in poverty so feared by climate-change contrarians. Within decades, a new energy paradigm will be enriching us and securing a future that works with nature, not against it.

## ACKNOWLEDGMENTS

Thanks are due my wife (and often coeditor) Pat Keiffer, my mother (and critical reader) Hazel Johansen, and Matthew Rothschild, editor-in-chief of *The Progressive*, all of whom pressed me to concentrate on solutions as well as problems. Editor Robert Hutchinson at Praeger added wit, wisdom, and wonderful suggestions. The idea for the book itself came from Anthony Chiffolo, editorial director at Praeger.

I also thank Clarinda Karpov of Omaha for materials on a "green" diet and global warming, as well as everyone who has helped, as they have for more than a quarter-century, at the University of Nebraska Criss Library. UNO's School of Communication director Jeremy Lipschultz, an accomplished author himself, has done his best to shield me from the pecked-to-death-by-ducks minutia of academia to free up time to research and for writing. Thanks as well to Dean Gail Baker of our College of Communication, Fine Arts, and Media.

## REFERENCES

*Business Week*. 2004. "How to Combat Global Warming: In the End, the Only Real Solution May Be New Energy Technologies." August 16, 108.

Canadell, Josep G., Corinne Le Quéré, Michael R. Raupach, Christopher B. Field, Erik T. Buitenhuis, Philippe Ciais, Thomas J. Conway, Nathan P. Gillett, R. A. Houghton, and Gregg Marland. 2007. "Contributions to Accelerating Atmospheric $CO_2$ Growth from Economic Activity, Carbon Intensity, and Efficiency of Natural Sinks." *Proceedings of the National Academy of Sciences*. Published online before print, October 25.

Environment News Service. 2007t. "Industrialized Countries' Greenhouse Gases Hit Record High." November 20. http://www.ens-newswire.com/ens/nov2007/2007-11-20-02.asp.

Havel, Vaclav. 2007. "Our Moral Footprint." *New York Times*, September 27. http://www.nytimes.com/2007/09/27/opinion/27havel.html.

McKibben, Bill. 2007. "The Race against Warming." *Washington Post*, September 29, A-19. http://www.washingtonpost.com/wp-dyn/content/article/2007/09/28/AR2007092801400_pf.html.

National Academy of Sciences. 1991. *Policy Implications of Greenhouse Warming*. Washington, D.C.: National Academy Press.

*Science*. 2007. "Melting Faster" [review of *Geophysical Research Letters* 34, L09501 (2007)]. Editors' Choice. 316 (May 18): 955.

Steinman, David. 2007. *Safe Trip to Eden: 10 Steps to Save Planet Earth from Global Warming Meltdown*. New York: Thunder's Mouth Press.

# Introduction

For the past two centuries, at an accelerating rate, the basic composition of the Earth's atmosphere has been materially altered by the fossil-fuel effluvia of machine culture. Human-induced warming of the Earth's climate is emerging as one of the major scientific, social, and economic issues of the twenty-first century, as the effects of climate change become evident in everyday life in locations as varied as small island nations of the Pacific Ocean and the shores of the Arctic Ocean.

This book, written for general readers in a broad range of audiences, describes what people can do as individuals to combat global warming. These individual actions are placed in a broader context of public policy, scientific debate, and technological innovation.

I have endeavored here to describe solutions for global warming at the personal, technological, and public-policy level, describing how each realm affects the others, with an emphasis on what people can do in their daily lives to reduce their "carbon footprints." This book is quite purposively a mixture of the prosaic, the necessary, the possible, and the improbable. Many of these changes are simple things we can do today, such as properly inflating a car's tires, using compact fluorescent light bulbs, driving less, and recycling (manufacture from recycled goods requires less energy than original manufacture). Others are wide-ranging innovations in energy infrastructure that will occur over decades—new ways of harvesting the sun with arrays of mirrors, for example. Quite a number of rather simplistic how-to books have or will be published in this field, but I find many lacking in connections between personal choices and legal and political context that would give them meaning. The last few years have brought us an ample cargo of books asserting that we can save the world by being "green" without much effort. Their titles are suggestive: *It's Easy Being Green* (Crissy Trask 2006); *The Lazy Environmentalist* (Josh Dorfman 2007); *The Green Book: The Everyday Guide to Saving the Planet One*

*Simple Step at a Time* (Elizabeth Rogers and Thomas M. Kostigen 2007). The real path to a sustainable world will be more of a challenge.

## OUTLINE OF THE BOOK

This is a book about necessary changes in the way we do business and live our lives. Our circumstances are begging for revolutionary basic modifications in the ways we obtain and use energy. While the technology of communication has changed during the last two decades via digitization in ways that our grandparents never would have imagined, our transportation and energy infrastructure is basically the same one that ushered in the fossil-fuel age a century and a half ago.

I begin, in the first chapters, with human endeavors that produce most of our carbon footprints—transportation (surface and air) and housing, including food. Today we can properly inflate our tires, share rides, and carpool. In the longer range, however, we will have to transform our sprawling cities so that we do not spend so much time traveling (and often sitting still) in cars.

The United States, with its vast distances and large, mobile population, has become a world leader in air travel, the most carbon-unfriendly way to get from place to place. A seat on an airplane routinely puts two to three times as much carbon into the air (among other pollutants) as travel by automobile. Trains are even better than cars, one reason they are being revived and improved around the world. Even Amtrak is coming aboard. Europe's older and more compact cities are already rearranging how people move, with trains and bicycles in the ascent. Some trains in Europe, China, and Japan now travel at more than 200 miles an hour.

"Green" building is on the rise around the world, in many surprising and innovative ways, and architectural engineers now ask how a building will affect the environment as a basic part of their planning. The amount of energy that can be shaved from our daily use by attending to conservation in homes and offices can be astounding—more than half, at times, using remodeling of older structures. Likewise, more attention is being paid to the distances traversed by our food before it reaches our tables. Food grown closer to home is not only cheaper to transport, reducing carbon emissions, but usually tastes better as well. Here, as elsewhere, the most practical course is often the most environmentally sensitive and sensible.

Chapters 4 through 7 follow with an examination of energy infrastructure changes (biofuel, wind, solar, and other options). Corn ethanol is at best a temporary solution because its manufacture requires almost as much energy as it yields. With its customary short-sightedness, however, the George W. Bush administration, for political reasons, has dived headlong into corn ethanol with massive subsidies. I live in Nebraska, where the ethanol boom has been good for many farmers who have long suffered from low land and crop prices, so I regret raining on the ethanol parade, but environmentally it makes very little sense to force competition between food and fuel. Once technological problems have been solved, switchgrass and inedible parts of crops will become better fuel sources, as will animal wastes, which are already being used in some Midwest localities.

In the meantime, other energy sources are booming. Geothermal is now routinely being installed when schools and other public buildings are renovated (I have been watching one such conversion at a Catholic grade school a few hundred yards from my office). Wind power, once ridiculed as pricey and undependable, has been vastly improved with new technology to the point that it is often cost-competitive with coal and oil. The story with solar is similar, although cost is still relatively high. Capacity for all these forms of energy is surging at unprecedented rates around the world. Chapter 7 considers new attention to older technologies, such as nuclear power, changes in land use (including farming), and "clean" coal.

Chapter 8 considers political and economic issues, as control of greenhouse gases has become newly popular in some formerly unlikely venues, from oil companies to Wal-Mart, as businesspeople realize that there is money to be made. How real is this drive toward green business, as public-relations people for the same company sometimes sing green as marketers and lobbyists cling to old habits?

Can a U.S. auto industry that lobbies against higher mileage standards really argue that it is "green" in Spin-Control Alley? Mileage for U.S. internal combustion engines has, in fact, declined during the last two decades, but gains in energy efficiency have been more than offset by increases in vehicle size, notably through sports-utility vehicles. And now comes a mass advertising campaign aimed at security-minded Americans for the biggest gas-guzzler of all: the fortress-like Hummer. Present-day automotive marketing may seem quaint in a hundred years. By the end of this century, and perhaps sooner than that, the internal combustion engine and the oil- (and natural-gas-) burning furnace will become museum pieces. They will be as antique as a horse and buggy is today. Such changes will be beneficial and necessary in the long range.

Are carbon offsets (donating to eco-constructive activities in the future while continuing to pollute today) real, or just a way to discharge guilt? Many corporations now have "sustainability officers," but how serious is corporate commitment? Can capitalism, which has been predicated on ceaseless growth and premeditated waste, really adjust to a new world in which more and bigger is not always better? Can we, in our life-guiding assumptions, unlearn a millennia-old ideology that instructs us to multiply and subdue the Earth?

Chapter 9 concludes with "technofixes," ideas that may or may not bear fruit, some of which are visionary (such as J. Craig Venter's search for a workaholic bacterium that will consume carbon dioxide), others silly (giant space mirrors to reflect sunlight) or potentially dangerous (filling the upper atmosphere with cooling, acidic sulfur, like a giant human-created volcano).

## GETTING THE MESSAGE

As temperatures rise, energy policy in the United States under the George W. Bush administration generally ignored atmospheric physics. By the time of this writing, however, even Bush was paying at least lip service to global warming's impact and the necessity of fashioning solutions.

Following the assumption of power by Democrats in the U.S. Congress in the 2006 elections, global warming also became a corporate and political priority in the United States—a far cry from the days when "sound science" in the Senate and the Oval Office meant inviting science-fiction writer Michael Crichton (author of *State of Fear*, et al.) to discourse on how greedy greenmailers were using the issue to make themselves rich and famous.

Even now, however, debate within a republic governed by special interests still outweighs actions that will become more necessary as the twenty-first century continues. Hot air reduces no greenhouse gases. Thermal inertia makes global warming a sneaky, backdoor emergency in which the atmospheric heat that we feel is already a half-century old on the greenhouse clock. Diplomacy and debate are ill-suited to a crisis in which the circumstances that compel us to action arrive only after our ecosystem-disabling actions are decades old. The public must realize what the scientists know—feedbacks involving melting sea ice (especially in the Arctic) and melting permafrost already have started us down a dangerous path.

## TOOLS ALREADY AT HAND

Solutions to global warming are not theoretical, for the most part. Many of the necessary tools are at hand. The city hall of Moab, Utah, for example, has been running 50 percent on wind power (for electricity) since 2003; the building is also heated and cooled using geothermal energy. In addition to employing alternative energy sources, the most effective way to avoid generating greenhouse gases is not to use a resource that would have been wasted (that is, by energy conservation). This alternative is often ignored in our capitalistic culture because it makes little or no money. Waste is built in.

So how, on a worldwide scale, do we cut greenhouse-gas emissions enough by midcentury to stabilize the world's climate? Fred Pearce, in *With Speech and Violence: Why Scientists Fear Tipping Points in Climate Change* (2007), sketched a set of solutions borrowed from the work of Robert Socolow, an engineer at Princeton University, to reduce the present 8.2 billion tons of human emissions to 2.2 billion, or about the 80 percent that many scientists say would address the problem:

1. Universally adopt efficient lighting and electrical appliances in homes and offices
2. Double the energy efficiency of 2 billion cars worldwide
3. Build compact urban areas served by efficient public transport, halving the use of automobiles worldwide
4. Expand worldwide capacity for wind power fifty times
5. Expand use of biofuels fifty times
6. Embark on a worldwide program to insulate buildings
7. Cover an area the size of New Jersey with solar panels

8. Quadruple current electricity production from coal to natural gas, converting coal-fired power stations
9. Capture and store carbon dioxide from 1,600 gigawatts of natural-gas power plants
10. Halt global deforestation and plant an area the size of India with new forests
11. Double nuclear-power capacity
12. Increase by tenfold global use of low-tillage farming methods to increase soil storage (Pearce 2007, 249)

## THE KEY: REDUCING POLLUTION FROM COAL-FIRED ELECTRICITY GENERATION

Emissions from 2,100 coal-fired power plants worldwide emit a third of humankind's carbon dioxide. In the United States, roughly 600 coal-powered plants produce 30 percent of the country's 7 million metric tons of greenhouse gases, as much as all the cars and all other industries combined. More than four-fifths of the carbon dioxide emissions from power generation in the United States come from coal (82.3 percent, versus 13.4 percent from natural gas and 4 percent from oil), according to the U.S. Department of Energy (Carmichael 2007, 4-B). In 2006, China generated 80 percent of its power with coal and was adding one new major generating plants every two weeks (Kintisch 2007b, 184).

Buildings are the single largest consumer of power from coal, three-quarters of the total. Reducing new and existing buildings' use of electricity 50 percent by 2030 would remove the need for more coal-fired power in the United States, according to Architecture 2030. This group of architects seeking solutions to global warming adds that building the 150 new coal plants now planned for the United States will make adverse climate change nearly impossible to avoid. The same goes even more emphatically for China.

James E. Hansen, director of NASA's Goddard Institute for Space Studies, has recommended a moratorium on construction of new coal-fired power plants until technology for carbon dioxide capture and sequestration is available (see chapter 7). This step alone (if it is adopted worldwide) may go a long way toward preventing descent into a runaway greenhouse effect in coming decades. About a quarter of power plants' carbon dioxide emissions will remain in the air "forever"—that is, more than 500 years—long after new technology is refined and deployed. As a result, Hansen expects that all power plants without adequate sequestration will be obsolete and slated for closure (or at least retrofitting) before midcentury (Hansen 2007b). The contribution of coal-fired power is so important to the worldwide carbon cycle that Hansen and Makiko Sato believe that "annual $CO_2$ emissions should begin to decline once phase-out of old-technology coal-fired power plants begins" (Hansen and Sato 2007).

Hansen believes that "coal will determine whether we continue to increase climate change or slow the human impact. Increased fossil fuel $CO_2$ in the air today, compared to the pre-industrial atmosphere, is due 50 per cent to coal, 35 per cent to oil and 15 per cent to [natural] gas. As oil resources peak, coal will determine future $CO_2$ levels" (Hansen 2007a).

Riding in the Iowa countryside during the spring of 2007, Hansen had a haunting thought:

> Recently, after giving a high-school commencement talk in my hometown, Denison, Iowa, I drove from Denison to Dunlap, where my parents are buried. For most of 20 miles there were trains parked, engine to caboose, half of the cars being filled with coal. If we cannot stop the building of more coal-fired power plants, those coal trains will be death trains—no less gruesome than [as] if they were boxcars headed to crematoria, loaded with uncountable irreplaceable species. (Hansen 2007a)

Today 645 coal-fired plants produce about half the United States' electricity. As recently as May 2007, more than 150 new ones were planned, as electricity demand rose at a 2.7 percent annualized rate. The only way to make these plants unnecessary is to fund ways to curtail electricity consumption as if the Earth's future depends on it. With serious conservation efforts and application of existing technology, we can do this. *The best kilowatt is the one we do not use.*

Many people have been realizing that building new coal-fired power plants without controlling their emissions of carbon dioxide is a climate trap. By mid-2007, several new coal-powered generating plants were being canceled or postponed across the United States, even as the federal government continues to subsidize construction of coal-fired power through rural cooperatives (see chapter 8). A private equity deal worth $32 billion involving the TXU Corporation trimmed eight of eleven planned coal plants, as similar plants were scuttled in Florida, North Carolina, Oregon, and other states.

A proposed a 750-megawatt coal-fired power plant in Waterloo, Iowa, was refused permission to annex residential land on which to build by a city development board during October 2007. One landowner whose home was in the path of the plant, Merle Bell, a retired farmer, stood in the way of the plant's developers. At one point during September 2007, more than 500 people rallied on his century-old farm in support, even as LS Power project manager Mark Milburn said that the plant could be built without the extra land. The question was not if it *could* be built, but whether it *should* be. Sierra Club organizer Mark Kresowik said: "This is the beginning of the end for LS Power in Iowa. States across the country are recognizing the dangers of overdependence on coal and have rejected new plants" (Environment News Service 2007j).

A week later, the Kansas Department of Health and Environment on October 18, 2007, became the first government agency in the United States to use carbon dioxide emissions as a reason to reject an air permit for planners of a coal-fired electricity generating plant. It rejected an application by Sunflower Electric Power, a rural electrical cooperative, to build two, 700-megawatt, coal-fired plants in Holcomb, in western Kansas, the site of the savage murders described by Truman Capote in his book *In Cold Blood* (Mufson 2007f).

About two dozen coal plants had been canceled by early 2007, according to the National Energy Technology Laboratory in Pittsburgh, an agency of the U.S. Department of Energy. Citibank downgraded the stocks of coal-mining companies in mid-July 2007, saying "prophesies of a new wave of coal-fired generation have vaporized" (Smith 2007b, A-1). Climate-change concerns are often cited when coal plants are canceled, especially in Florida, where rising sea levels from melting ice in the Arctic, Antarctic, and mountain glaciers is already eroding coastlines. Florida's Public Service Commission is now legally required to give preference to alternative energy projects over new fossil-fuel generation of electricity. The states of Washington and California have been moving toward similar requirements. Xcel Energy and Public Service of Colorado were allowed to go ahead with a 750-megawatt coal-fired power plant only after they agreed to obtain 775 megawatts of wind power.

By the beginning of 2008, the European Commission was weighing whether to require new power stations to plan space to retrofit in a way that will store greenhouse-gas emissions via carbon-dioxide capture and storage (CCS) technology, the first legal move of this type in the world. If the European Parliament and Council approve this proposal it could become law in 2009, a large step toward making CCS a commercial reality. The requirement as written does not contain a date on which actual CCS would be required. Installation of CCS technology, now still in its infancy, could reduce global $CO_2$ emissions by one-third by 2050, if widely deployed (Schiermeier 2008, 232). At present, Norway, Britain, China, and the United States are planning CCS pilot plants.

Legislation was introduced in March 2008 in the U.S. House of Representatives to require new coal-fired electric generating plants' use of state-of-the-art CCS technology. Rep. Henry Waxman, Democrat of California, chair of the House Government Oversight Committee, and Rep. Edward Markey of Massachusetts, chair of the House Select Committee on Energy Independence and Global Warming, introduced the bill (the Moratorium on Uncontrolled Power Plants Act of 2008.) The bill places a moratorium on permits for new coal-fired power plants without CCS from the U.S. Environmental Protection Agency or state agencies ("National Ban" 2008).

At about the same time, several groups rallied about 100 people at the Kansas State Capitol, March 11, in support of Gov. Kathleen Sebelius, who was expected to veto an energy bill that would have permitted two new 700-megawatt coal-fired power plants without CCS in the western part of that state, near Holcomb ("Kansans" 2008). The bill had narrowly passed the state House and Senate after Rod Bremby, secretary of the Kansas Department of Health and Environment, denied a permit during 2007

("Kansans" 2008). The bill also sought to eliminate the department's role in issuing permits for power plants.

Environmental groups tightened their focus on proposed coal-fired power plants during 2007. Environmental Defense and the Natural Resources Defense Council assembled "strike forces" to mobilize opposition to new plants state by state. These strike forces played a role in obtaining cancellation of plants in Florida and Texas. In New Mexico, for example, the groups intervened in a dispute over whether to construct a new power plant on the Navajo Reservation, where the state government, which is opposed to the project, has no direct power to prevent it. The plant's carbon footprint would equal 1.5 million average automobiles. Coal-fired electricity contributes more than half of the 57 million tons of annual carbon dioxide emissions in New Mexico. Together, two existing plants in that state emit 29 million tons (Barringer 2007).

"Hey!" I hear someone exclaiming. "We *need* those new power plants so we won't freeze in the dark!" No, we don't. Without breaking a sweat, we can eliminate enough energy demand to make every one of them unnecessary, without sacrificing anything except a few million dollars on our collective electric bills. Read on to find out how. Here's a tease: I have become a conscious light-switch guerrilla, at home and at the office. I learned quickly how to douse several hundred excess hallway lights in my university office building on weekends and holidays. Call me the kilowatt killer. No one seems to have missed any of them.

## REFERENCES

Barringer, Felicity. 2007. "Navajos and Environmentalists Split on Power Plant." *New York Times*, July 27. http://www.nytimes.com/2007/07/27/us/27navajo.html.

Carmichael, Bobby. 2007. "Opposition Takes on Coal Plants." *USA Today*, October 30, 4-B.

Environment News Service. 2007j. "Coal-fired Power Plant Blocked in Iowa." October 15. http://www.ens-newswire.com/ens/oct2007/2007-10-15-093.asp.

Hansen, James E. 2007a. "Coal Trains of Death." James Hansen's E-mail List, July 23.

———. 2007b. "Political Interference with Government Climate Change Science." Testimony before the House Committee on Oversight and Government Reform, 110th Cong., 1st sess., March 19.

Hansen, James, and Makiko Sato. 2007. "Global Warming: East-West Connections." Unpublished.

"Kansans Rallied to Resist Coal-Burning Power Plants." 2008. Environment News Service, March 12. http://www.ens-newswire.com/ens/mar2008/2008-03-12-092.asp.

Kintisch, Eli. 2007b. "Making Dirty Coal Plants Cleaner." *Science* 317 (July 13): 184–86.

Mufson, Steven. 2007f. "Power Plant Rejected over Carbon Dioxide for First Time." *Washington Post*, October 19, A-1. http://www.washingtonpost.com/wp-dyn/content/article/2007/10/18/AR2007101802452_pf.html.

"National Ban on New Power Plants Without $CO_2$ Controls Proposed." Environment News Service, March 12, 2008. http://www.ens-newswire.com/ens/mar2008/2008-03-12-091.asp.

Pearce, Fred. 2007. *With Speech and Violence: Why Scientists Fear Tipping Points in Climate Change*. Boston: Beacon Press.

Schiermeier, Quirin. 2008. "Europe to Capture Carbon." *Nature* 451(January 17): 232.

Smith, Rebecca. 2007b. "New Plants Fueled by Coal Are Put on Hold." *Wall Street Journal*, July 25, A-1, A-10.

CHAPTER 1

# Our Cars Are Killing Us: Options in Personal Transport

I am definitely a commuting renegade. I am reminded of my status every day when I ride a bicycle or walk the two miles between my home and my office at the University of Nebraska at Omaha. Unless the temperature is well below zero with a stiff breeze (if the wind is in my face, I ask my wife for a ride when the wind chill hits 20 below), my commute consumes not an ounce of fossil fuels. It also benefits my health. I am definitely a small fish in a sea of automotive-enhanced urban sprawl, however. How I cherish brief memories of Copenhagen and Amsterdam, where bicycles have reached critical mass—a third of the commuting workforce or more—enough to pester drivers of motorized behemoths that are heavily taxed. In my real world, I am a mouse among elephants.

Cars are fast and easy, certainly, but they also are killing us by filling the atmosphere with greenhouse gases and turning many of us into nearly immobile blimps. Cars are an addiction: they shape their hosts to meet their needs. Most urban areas in the United States have been shaped by the automobile to such an extent that, for most people, other transportation options are unavailable, unappealing, and virtually impossible. To combat global warming, we will have to reshape our cities. The ultimate solution is to work at home (cutting commute time and energy consumption to zero), a solution that is becoming more appealing. My publishers now comprise virtual networks with people across the country, many of whom work at home, linked by digital technology. The next-best option is to live close enough to the office to sharply reduce or eliminate one's commuting carbon footprint. University faculties need classrooms, meeting spaces, and offices, but we *can* live close to our worksites.

The automobile is an exemplar of fossil-fuel culture—easy to use, convenient, suited to the individualism of our time, and generative of greenhouse gases in its manufacture, as well as its everyday use. Each auto-mobile brings with it a bundle of carbon dioxide, carbon monoxide, meth-ane, nitrous oxide, and waste heat. The tailpipes of motorized vehicles emit 47 percent of North America's nitrous oxides, a primary component of ozone in the lower atmosphere (Gordon and Suzuki 1991, 198). The automobile, in addition to burning fossil fuels, is also a little greenhouse factory of its own, as the windshield captures the sun's heat and cooks the air inside its steel-and-glass casing.

## CITIES SHAPED BY CARS

To a degree unknown elsewhere in the world, cities in the United States have been remade in the image of the automobile. Any U.S. city (or suburb) that reached maturity after the 1930s—the last decade in which the average person did not own a car—has become an energy-gobbling sprawling mass of suburbs and freeways that virtually guarantees that most people need automobiles whether they want them or not.

The suburban world in the United States has been shaped so acutely by the automobile that anyone without one may have trouble getting around. Witness Leesburg, Virginia, resident Jose Vetura, who could walk from home to his restaurant job in ten minutes—except that a six-foot iron fence and six lanes of traffic stand in his way. According to an account in the *Washington Post*, Vetura

> follows a well-worn dirt path along the fence, past fast-food wrappers and dandelions to the intersection of Edwards Ferry Road and the Route 15 bypass. There, surrounded by shards of glass and cigarette butts, he waits for a gap in the traffic and bolts across the bypass without any help from a cross-walk. After hiking up a gravel hill, he crosses a vast parking lot past Ruby Tuesday and Ross Dress for Less before reaching his destination in about 20 minutes. (Chandler 2007, B-1)

Especially in the suburbs, anyone who walks any distance at all does so in harm's way. Overpasses, sidewalks, crosswalks, and bicycle paths have become a second thought—if that—in suburban planning. Outside of a few major eastern cities (New York and Boston being prominent exam-ples), the automobile has become an everyday necessity for nearly every-one. Weaning ourselves from global warming in the long run is going to require major surgery on our concepts of urban land use.

In some places, such surgery is being performed as we watch. Portland, Oregon, has canceled several freeway projects that once threatened to make the center city little more than a conduit to the suburbs. Even so, Portland has no shortage of downtown freeways, but it also has a well-used light-rail system that connects neighborhoods and its airport with the central core. Property values have risen in the center city as people reno-vate homes that are within bicycling or walking distance of downtown businesses. Even in Detroit, General Motors during 2007 entered a part-nership to develop residences in once-abandoned buildings along the waterfront. One of the selling points for these lofts is—note that the pitch is being made by an automobile manufacturer—the fact that people who live in them can *walk* to work at GM's headquarters (White 2006, D-4).

Such experiments, to date, have been exceptions, however. Even as gas prices soared, the climate warmed, and public concern over climate change increased, most people in the United States complained but filled their gas tanks like slaves. The amount of time spent stalled on congested highways continued to increase. A commuter in Los Angeles spent 72 hours a year wasting time and gasoline during 2006; in most other large U.S. cities, the average was 50 to 60 hours, according to the U.S. Department of Transportation. Gasoline demand proved notoriously inelastic (as economists say) in the face of rising prices. What choice did most people have? Living in Omaha, a very average U.S. city, I could see the point that my bicycle-riding habits (and my insistence that expensive gas might be good for all of us in the long run) were, as an early reader of my work told me, a bit naïve and elitist. I could afford a house within two miles of my office (a rarity these days, it seems); given rain or snow, I could walk if necessary. Omaha has no subway, and even public buses are very scarce. I stood at the edge of my campus on Dodge Street, a main Omaha artery, with its stream of cars and trucks, remembering the main arterial that bordered the Catholic University of Lublin, Poland, where I taught in 2005, with its thick foot traffic and packed buses every minute or two.

"What about the woman who cleans houses for a living? She certainly wouldn't be able to afford housing in the neighborhoods where she cleans houses," my reader told me. I might have made a case for telecommuting, but maids can't mop floors over the Internet. Telecommuting is a white-collar option. As I recalled Europe, with its compact cities and abundant trains and buses, I was reminded that this is America, where car culture has killed all other forms of transport nearly everywhere. To paraphrase one-time defense secretary Donald Rumsfeld in another context, you go to war with the army you've got. Tomorrow, most American workers will have to suck it up and endure high fuel prices and urban sprawl in a land where freedom of choice means big car or small car, Coke or Pepsi, McDonald's or Burger King.

## AUTOMOBILE TYRANNY TIGHTENS

The average person in the United States' travel mileage has increased 60 percent in thirty years. From 1990 to 2004, Americans increased the number of miles driven annually by about 34 percent. Some 225 million cars were on the road in the United States in 2006, 40 percent of the world's total. Since 1970, car mileage in the United States per person has increased 50 percent, with lengthening commutes and longer vacations being major factors. Public transportation in the United States has fallen to 1 percent of total transport miles. Walking comprises a miserly 0.7 percent (Hillman, Fawcett, and Rajan 2007, 49–51, 53).

Between 1990 and 2003, greenhouse-gas emissions from passenger vehicles increased by 19 percent in the United States, caused mainly by increased sales of light-duty vehicles (SUVs, minivans, etc.) and an increase in the number of miles Americans travel every year, according to the U.S. Environmental Protection Agency. Meanwhile, the fuel economy of new vehicles has declined because of increasing sales of light-duty trucks, which overtook sales of passenger cars in 2002.

According to Bill McKibben, "The average American car, driven the average American distance—ten thousand miles—in an average ... year releases its own weight in carbon into the atmosphere" (1989, 6). Global warming, in a sense, is the exhaust pipe of the "American Dream" of the bigger and better, the new and improved, the mobile life in the fast lane. Cars and trucks used in the United States burn 15 percent of the world's oil production. Transportation alone consumes one-fourth of the energy and two-thirds of oil used in the United States (Cline 1992, 200).

## COMBATING CAR CULTURE IN EUROPE

In the meantime, efficiency standards for cars in Europe now are measured not in kilometers per liter, but in grams of carbon dioxide emitted per kilometer. A good car emits about 100; a big Volvo, more than twice as much. Sweden has a carbon tax, but it also gives tax credits for driving large domestic Volvos and Saabs. The Swedish carmakers have been resisting regulation, asserting that they are working on fuel efficiency and biofuels. At Saab, more than 80 percent of sales are now "biocars" (aka "flexfuel," using gasoline or ethanol). The Swedish government recently adopted a vehicle tax based on carbon dioxide emissions rather than weight. Per Bolund, a Green Party representative, pointed out that some cars such as hybrids are heavy but relatively low in fossil-fuel emissions.

To reduce oil consumption and greenhouse-gas emissions, Swedes are being encouraged to avoid commuting altogether by teleconferencing and using the Internet from home. When they do commute, more Swedes now use public transport, hybrid vehicles, and bio-diesel cars, as well as bicycles. Stockholm plans to introduce a fleet of electric hybrid buses in its public transport system on a trial basis during 2008. These buses will use ethanol-powered internal combustion engines and electric motors, an interim step toward development of entirely "clean" vehicles. The vehicles' diesel engines use ethanol.

Ulf Perbo, who heads BIL Sweden, the national association for the automobile industry, said that even automakers there want to end oil dependency. "It is not in our interest to be dependent on oil, with regard to the production and sales of cars. Oil is not what interests us; cars are. And oil is going to be a limitation [to the production and sales of cars] in the future" (Johansen 2007, 24). All Swedish gas stations are required by an act of Parliament to offer at least one alternative fuel. Every fifth car in Stockholm now drives at least partially on alternative fuels, mostly ethanol.

Closer to home, growing numbers of sports-utility vehicles roaming the roads of the United States recall the words of Henry Ford: "Mini-cars make mini-profits" (Commoner 1990, 80). His wisdom proved to be rather obsolete as gasoline prices rose to record levels in 2008, and the sales of U.S.-made "maxi" cars plummeted. Soon, Toyota passed General Motors as the largest auto company on the planet, and, even then, U.S. auto executives complained that enforcing higher mileage standards might put them out of business. Actually, reliance on maxi-cars had nearly done that already. By 2007, parts of the auto companies were beginning to realize this, just as the U.S. Congress sought to raise mileage standards in the United States to 35 miles per gallon (mpg) by 2020, a much lower standard than most other countries. The Toyota Prius hybrid was getting 50 miles per gallon of gasoline, and Toyota was selling 250,000 hybrids a year in the United States by 2006, when U.S. companies began to get the message.

If we can't give up our cars, could we at least downsize them? In the United States, 70 percent of car and light truck sales in 2006 had six- or eight-cylinder engines. In Europe, 89 percent had four cylinders or fewer (Spector 2007, B-1). Could our prototypical cleaning woman get to work on a couple fewer cylinders? Could soccer moms transport the kids on four cylinders? Even with a majority of people telling pollsters that global warming is a major problem, the outer exurban rings of our largest cities continue to expand. The sweet, easy, devastating tyranny of the individual automobile tightens its grip in the land of the free and the home of the brave.

## SIMPLE STEPS TO IMPROVE EFFICIENCY—NOW

Given the cars (and the cities) that we have, how can we reduce fuel consumption now? Some carbon dioxide emissions can be neutralized in ridiculously easy ways. If you can boost your gas mileage from 20 mpg to 24, your car will put 200 fewer pounds of $CO_2$ into the atmosphere each year (Bjerklie 2007). Fill you car's tires. A few pounds per square inch can be worth 3 percent in fuel efficiency by itself. Keeping a car in tune can improve mileage 5 percent. Correcting a serious maintenance problem, such as a faulty oxygen sensor, can improve mileage by as much as 40 percent. Replacing a clogged air filter can improve mileage 10 percent. Aggressive, "heavy-footed" driving can cost 5 percent (in town) to 33 percent (on the highway). Driving too fast on the highway (more than 55 to 60 miles per hour [mph]) taxes efficiency severely; every 5 mph over 60 reduces mileage 5 to 8 percent. Driving 50 mph saves 30 percent on fuel vis-à-vis 70 mph. Carrying an extra 100 pounds in a car costs 2 percent worth of mileage, so empty your car of unneeded weight. Combine errands and do them in a circle. Avoid rush periods in traffic.

Another easy carbon killer is to develop a preference for right turns over lefts. During 2004, United Parcel Service (UPS) announced that its drivers would avoid making left turns. The time spent idling while waiting to turn against oncoming traffic burns fuel and costs millions of dollars each year. A software program maps a customized route for every driver to minimize left turns. In metropolitan New York City, UPS reduced $CO_2$ emissions by 1,000 metric tons between January and April 2007 with such simple strategies (Sayre 2007).

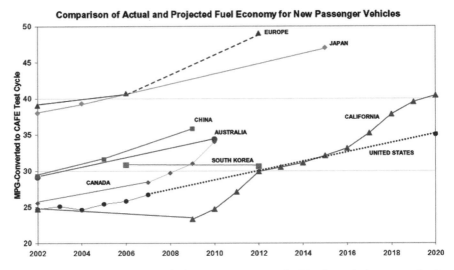

**Figure 1.1. Comparison of fuel economy and GHG emission standards (see chart).**
Source: Pew Center on Global Climate Change.

Consider carpooling. Share the ride with one person and cut your per-person carbon dioxide emissions in half. Carpool with three people and cut your share of emissions to a quarter of driving alone. Washington State used its Clean Air Act to provide businesses with state tax credits that encourage employees to share rides, drive shorter distances, and work longer hours on fewer days, reducing the number of commuting trips. The number of vehicle trips per day fell by 20,000 after the program started in 1991, saving commuters $13.7 million and 5.8 million gallons of gas, and reducing 78,000 tons of greenhouse gases (Masters 2007b).

## MANIFEST INEFFICIENCY

Our cars are monuments to fuel inefficiency. The average fuel economy of vehicles sold in the United States has remained nearly stagnant, at around 20 mpg, for several decades. Only 13 percent of a car's energy reaches the wheels, and only half of that actually propels the car. The rest is lost to idling, heat, vibration, and such accessories as air conditioning. Because 6 percent of a car's remaining energy converts to brake heating when it stops, less than 1 percent of the energy the car consumes ends up propelling the driver (Lovins 2005, 74, 76, 82–83).

The X Prize Foundation is sponsoring an automotive contest, with a prize of more than $10 million, to accelerate research leading to a car that can travel 100 miles on a gallon of gasoline. The same group earlier awarded $10 million to a team that built the first private spacecraft to leave the Earth's atmosphere (Bunkley 2007b). According to contest rules, the winning design must be commercially viable and production ready, not a "concept car" such as some presented at annual auto shows. Each team must prepare a business plan for building at least 10,000 of its vehicles at a cost comparable to that of cars now available (Bunkley 2007b). "The industry is stuck, and we think a prize is perfect to disrupt that dynamic," said Mark Goodstein, executive director of the Automotive X Prize. "Failure is frowned upon in this industry, and that doesn't make for big advances. It makes for incrementalism" (Bunkley 2007b).

"Several cars have been built that could travel more than 100 miles on a gallon, but they were expensive and were used only for demonstration. Building one that can go 100 miles per gallon, I think any of the automakers could do that," said James A. Croce, chief executive of NextEnergy, a nonprofit organization in Detroit that promotes alternative energy. "It's mass-producing them that's the problem" (Bunkley 2007b). "This is not a question of curing cancer," Goodstein said. "The technologies to build super-efficient vehicles exist. It's just a matter of convincing manufacturers to build them" (Bunkley 2007b).

## AUTOMOBILE EFFICIENCY: U.S. STATES TAKE ACTION

The California Air Resources Board (CARB), defying the auto industry, voted unanimously during late September 2004 to approve stringent rules reducing automobile emissions of greenhouse gases. Under the regulations,

**Figure 1.2. Traffic jam in Los Angeles, California. Auto traffic and urban sprawl continue to increase in the United States even as temperatures rise. U.S. Transportation Department.**
Courtesy of Shutterstock.

the automobile industry must cut exhaust from cars and light trucks by 25 percent and from larger trucks and sports-utility vehicles by 18 percent. The industry will have until 2009 to begin introducing cleaner technology and until 2016 to meet the new exhaust standards.

California's plan could eventually lead most of states on the East and West Coasts of the United States to require similar emissions cuts because the state's share of the national automobile market is so large. In turn, these requirements may provoke the automakers to adopt the same standards for cleaner, more fuel-efficient vehicles across their model lines. The only way to cut global warming emissions from cars is to use less fossil fuel. Because of this limitation, proposed cuts in legally allowable emissions would, as a side effect, force automakers to increase fuel economy by roughly 35 to 45 percent. California's plan, as proposed in 2004, requires automakers to cut greenhouse-gas emissions from their new vehicles by 29.2 percent over a decade, phasing in gradually from the 2009 to the 2015 model years.

During 2004, the governments of New Jersey, Rhode Island, and Connecticut said that they intended to follow California's automobile rules instead of the federal government's. New York, Massachusetts, Vermont, and Maine already had adopted the California rules. "Let's work to reduce greenhouse gases by adopting the carbon dioxide emission standards for motor vehicles which were recently proposed by the State of California," New York governor George E. Pataki said in his State of the State address in 2003. These seven states and California account for almost 26 percent of the U.S. auto market, according to R. L. Polk, a company that tracks automobile registrations (Hakim 2004, C-4). Automakers from Detroit to Tokyo believe that these states, along with Canada, could form a potent bloc on automobile regulation to cut emissions of greenhouse gasses. "It would be a logistical and engineering challenge, and a costly problem," said Dave Barthmuss, a spokesman for General Motors. "It's more cost-effective for us to have one set of emissions everywhere" (Hakim 2004, C-4). "If they only want to make one car," said Roland Hwang, a senior policy analyst at the Natural Resources Defense Council, "clearly it should be a clean car, and that's the California car" (Hakim 2004, C-4).

During 2007, the CARB voted to require automobile manufacturers to affix labels listing their vehicles' smog and greenhouse-gas emissions so that the state's two million buyers a year can compare them. "This simple tool will empower consumers to choose vehicles that help the environment," said CARB chairman Robert Sawyer. "Most Californians recognize climate change as a very serious problem. This label will help consumers make informed choices" (Environment News Service 2007h). The 25 million vehicles on California roads, which travel roughly 900 million miles a day, emit about 350,000 tons of greenhouse gases.

## WHERE THE RUBBER MEETS THE ROAD

Where the short-term profit-and-loss rubber met the public-relations road, the U.S. automobile industry seemed by 2007 to be speaking with more than one voice on the mileage-standard issue. On May 31, 2007, Ron Gettelfinger, the president of the United Automobile Workers union, and William Clay Ford Jr., executive chairman of the Ford Motor Company, told Michigan business leaders that cutting emissions and raising fuel-economy standards are critical to the future of the industry and that of Michigan. Gettelfinger said that "climate change is real" and that consumers are searching for more environmentally friendly vehicles. "If the auto industry continues to be seen as dragging its feet on environmental issues," he said, "it's going to hurt our brands and vehicles in the marketplace" (Bunkley 2007a).

Along with many other corporations, U.S. automakers had picked up the "green" mantra, endorsing greenhouse-gas limits in theory. When the debate devolved to practice, however, the same companies complained that enforcing meaningful mileage increases would deprive people of jobs and the companies of profits—this from companies that had been swimming in red ink after years of slavishly manufacturing large, inefficient gas-guzzlers as gasoline prices rose and more nimble companies in other countries (many of them Japanese) gobbled market share, even in the United States. Between 1984 and 2002, the weight of the average automobile in the United States gained 20 percent as mileage stagnated—after a 62 percent increase in gas mileage between 1975 (when the first legal limits on mileage at the federal level were passed) and 1984 (Surowieki 2007, 25). The automakers' dual corporate personalities endured.

The well-oiled pursuit of special interests was on display as the U.S. Congress debated energy legislation during June 2007. The Senate on June 21, 2007, dealt domestic car manufacturers a major blow by passing a bill requiring a major increase in mileage standards (from 25 mpg to 35 mpg by 2020), the first in two decades. The vote, 65 to 27, sent the bill to the House. Raising the mileage standard 10 mpg will reduce U.S. oil consumption by 1.2 million barrels a day and reduce emissions of greenhouse gases by an amount equal to removing 30 million cars from the road. Getting really serious about mileage, raising the standard to 55 mpg by 2020—a target for which technology is now available—would cut oil demand for transportation in the United States by half (Steinman 2007, 239).

Within days of the Senate's action, a full-page advertisement ran in the *Omaha World-Herald* (and a number of other U.S. newspapers) under the sponsorship of auto dealers urging voters to contact their congressional representatives to oppose what they called "extreme" mileage standards, such as the 35 mpg called for in Senate legislation.

By the end of 2007, however, congressional negotiators neared agreement on an energy bill that would raise average mileage of U.S. automobiles to 35 mpg by 2020. The auto industry agreed to the mileage standards for passenger cars, but convinced legislators to lower standards for some vehicles, most notably "light trucks" (e.g., SUVs). The bill, which also requires large increases in use of biofuels (13 billion gallons a year by 2012 and 36 billion gallons by 2022), offers a fuel-efficiency credit for flex-fuel vehicles. The bill requires that feedstocks other than corn be used in escalating amounts after 2013.

While it opposed practical legal requirements to bring down greenhouse-gas levels, General Motors bragged about its green consciousness in a press release: "GM is very pleased to join ... to proactively address the concerns posed by climate change and applauds its members for recognizing the important role that technology can play in achieving an economy-wide solution," said Richard Wagoner, chairman and chief executive officer (CEO) of General Motors (Environment News Service 2007w).

At the same time, however, Republican senators blocked a Democratic plan to raise taxes on oil companies by about $32 billion, with the money to be directed into tax breaks for wind power, solar power, ethanol, and other renewable fuels. Many of the same senators also blocked a law that would have required electric utilities to substantially raise the proportion of power they get from renewable energy to 15 percent by 2020. Thus, even at a time when green seemed to have become everyone's favorite color (at least for public-relations purposes), old lobbying habits died hard. A *New York Times* report noted that automakers' opposition to the mileage standard had been "ferocious" and added, "The clashes and impasses also provided a harbinger of potentially bigger obstacles when Democrats try to pass legislation this fall to reduce emissions of greenhouse gases tied to global warming" (Andrews 2007b).

Within days after their ferocious battle against stiffer mileage standards on Capitol Hill, Ford Motor Company and the Chrysler Group announced at a press conference in Washington, D.C., that they had joined the United States Climate Action Partnership (USCAP), a coalition lobbying for national legislation to impose legally binding limits on global-warming emissions. The partnership includes twenty-three of the world's largest corporations and six of the United States' best-known environmental groups. General Motors already had joined—even before the battle over mileage standards. "We are pleased to join USCAP at a critical stage in the conversation on climate change, energy consumption and environmental protection," said Alan Mulally, president and CEO of Ford (Environment News Service 2007b). No explanation was forthcoming as to why the same automakers seemed tooth-and-nail opposed to practical solutions to warming.

"We all recognize it is time for action," said Mulally. "Ford has long supported the six major principles of USCAP, including recognizing the importance of technology and the need to be environmentally effective. We have been actively developing a range of advanced technology vehicles to address the climate change issue. Now is the time for advancing a national approach to climate change where all of us—individuals, industry and government—take action toward reducing emissions of greenhouse gases," said Tom LaSorda, president and CEO, Chrysler Group (Environment News Service 2007b). Observers were left shaking their heads and asking, "Will the real automakers please stand up?"

Three months before their tenacious assault on higher gas-mileage standards, during March 2007, the chief executives of the same automobile companies had pledged to support mandatory caps on carbon emissions, as long as the caps covered all sectors of the economy. The CEOs delivered their promises to a House committee run by Rep. John Dingell—the crusty Michigan Democrat who, in his pre-green days, had favored whatever the automobile companies wanted. By 2007, Dingell was even sponsoring legislation for an energy tax of a type widely used in Europe, while stating that his old allies in the auto industry probably would lobby it into oblivion.

During the same year, a group of institutional investors, the largest of which was Merrill Lynch, called for mandatory emissions curbs of 60 to 90 percent by midcentury. This group called for companies to disclose investor risks associated with climate change to the Securities and Exchange Commission. Carl Pope, executive director of the Sierra Club, said that the political landscape on global warming was changing so quickly in 2007 that even he was unsettled: "I've rarely thought things were moving too fast to keep up with, and now I think things are moving too fast to keep up with" (Hitt 2007, A-6).

## A VOLVO FACTORY GOES CARBON DIOXIDE-FREE

While U.S. automakers were dickering over mileage standards, during September 2007 the Volvo Trucks plant in Ghent, Belgium, which employs 2,500 people and produces about 35,000 vehicles a year, became the first vehicle manufacturing plant in the world to operate without emitting any carbon dioxide. "Our ambition is to make all our plants $CO_2$-free and Ghent is the first," said Volvo CEO Leif Johansson. "It is not an easy undertaking, but we are prepared to try different alternatives to achieve our goal for $CO_2$-free production in our plants" (Environment News Service 2007cc). Volvo has production facilities in eighteen countries.

How does the factory do it? The Ghent factory has three wind-power generating stations on its site, producing half of its electricity. The rest is certified green energy supplied by the Belgian energy company Electrabel. A pellet-fired biomass plant supplies 70 percent of the heating requirements for the Ghent plant, while energy for combustion processes comes from solar cells on the roof. The remaining 30 percent is delivered by a former oil-fired boiler converted to burn biomass fuel.

## HYDROGEN FUEL: HYPE AND REALITY

Political correctness vis-à-vis global warming in the automobile industry has become associated with development of hydrogen fuel cells, especially after President George W. Bush used his State of the Union Address in January 2003 to propose $1.2 billion in research funding to develop hydrogen-fuel technologies. With those funds, Bush said that America could lead the world in developing clean, hydrogen-powered automobiles.

Jeremy Rifkin, a liberal social critic and author, published a book during September 2002 titled *The Hydrogen Economy: The Creation of the*

*Worldwide Energy Web and the Redistribution of Power on Earth.* Rifkin received the idea of a "hydrogen economy" nearly as manna from heaven. He believes that cheap hydrogen power could make the twenty-first century more democratic and decentralized, contrasting with oil's role during the nineteenth and twentieth centuries in the rise of powerful corporations and large nation-states. With hydrogen, wrote Rifkin, "Every human being on Earth could be 'empowered'" (Coy 2002, 83).

The power of the presidency to focus attention and resources on an issue—even prematurely—is amazing. Very suddenly, a hydrogen rush was on. On September 5, 2002, a coalition of companies, including DuPont and 3M asked Congress to spend $5.5 billion during the ensuing decade to advance fuel-cell development (Coy 2002, 83). During 2002, General Motors CEO Wagoner launched a $500 million hydrogen-car initiative for an anticipated (and later postponed) initial rollout in 2008. The cost could rise to billions of dollars annually if GM chooses to mass-produce the new models. Wagoner defended the expense as a crucial investment in GM's future: "People say, 'How can you afford to spend so much on fuel cells?' and I say, 'How can you afford not to?'" (Lippert 2002, E-3).

At the Paris Motor Show during September 2002, General Motors unveiled the Hy-wire, a "concept car" that uses a hydrogen fuel cell, about six years behind Toyota and Honda. Ford has purchased a sizable stake in British Columbia's Ballard Power, a world leader in the race to provide the first commercially available hydrogen fuel-celled automobile. GM also owns a stake in this company.

## NO FREE CLIMATIC LUNCH

Hydrogen fuel has been hailed as a godsend. Imagine a car that emits nothing but a little water vapor. It is a thought worthy of high praise, but imagination runs ahead of technology's ability to deliver. As much as it has been touted as pollution-free, hydrogen fuel is no free climatic lunch. Despite surface appearances, hydrogen manufactured with today's technology is no cleaner a fuel than gasoline.

The hype surrounding a hydrogen-based economy has been woefully premature, for one very important technological reason. Hydrogen, unlike oil or coal, does not exist in nature in a combustible form. Hydrogen is usually bonded with other chemical elements, and stripping them away to produce the pure hydrogen necessary to power a fuel cell requires large amounts of energy. Unless an alternative source (such as Iceland's geothermal resource) is available, hydrogen fuel usually is produced from fossil fuels. Extraction of hydrogen from water via electrolysis and compression of the hydrogen to fit inside a tank that can be used in an automobile requires a great deal of electricity. Until electricity is routinely produced via solar, wind, and other renewable sources, the hydrogen car will require energy from conventional sources, including fossil fuels. Today, 97 percent of the hydrogen produced in the United States comes from processes that involve the burning of fossil fuels, including oil, natural gas, and coal.

Paul M. Grant, writing in *Nature*, provided an illustration:

> Let us assume that hydrogen is obtained by "splitting" water with electricity—electrolysis. Although this isn't the cheapest industrial approach to "make" hydrogen, it illustrates the tremendous production scale involved—about 400 gigawatts of continuously available electric power generation [would] have to be added to the grid, nearly doubling the present U.S. national average power capacity. (Grant 2003, 129–30)

That, calculated Grant, would represent the power-generating capacity of 200 Hoover Dams. At $1,000 per kilowatt, the cost of such new infrastructure would total about $400 billion. What about producing the 400 gigawatts with renewable energy? Grant estimated that, "with the wind blowing hardest, and the sun shining brightest," wind power generation would require a land area the size of New York State or a layout of state-of-the-art photovoltaic solar cells half the size of Denmark (2003, 130). Grant's preferred solution to this problem is use of energy generated by nuclear fission.

Another problem with hydrogen fuel is storage, in both vehicles and fueling stations. Hydrogen is flammable (far more prone to explode than gasoline) and must be stored at high pressure (up to 10,000 pounds per square inch). It is also far less dense than conventional fossil fuels and so requires fifty times the storage space of gasoline. Liquid hydrogen avoids these problems, but it must be stored at –400°F, not a practical solution for Everyman's car or every neighborhood's fueling station. In addition, delivering air chilled to –400°F would itself involve a great deal of energy (meaning expense and fossil-fuel emissions). Nevertheless, the Energy Department by 2007 was pouring grant money ($170 million over five years) into developing a fleet of fuel-cell vehicles and fueling stations (*Houston Chronicle* 2007, 2-D).

Buzzworthy energy technologies came and went at the Bush White House. One day it was hydrogen, the next it was corn ethanol—an enormous problem with energy policy that was governed by spin-control, with scant attention to (or knowledge of) the real world and its challenges. No one at the White House seemed to realize that, with present technology that requires a fuel cell, the cost of hydrogen energy is much higher than conventional fuels. As of 2007, hydrogen fuel cost the equivalent of $6–10 a gallon for the energy delivered by a gallon of gasoline. Thus, without several major scientific breakthroughs, hydrogen cars are a pipe dream (Romm 2007, 186–87).

## HYDROGEN FUEL A SUCCESS IN ICELAND

While hydrogen isn't the magic worldwide solution that some of its proponents imagine, premonitions of hydrogen-based transportation systems have been emerging in small ways. In Iceland, for example, 85 percent of the country's 290,000 people use geothermal energy to heat their homes (Brown 2006, 166). Iceland's government, working with Royal Dutch Shell and Daimler-Chrysler, in 2003 began to convert Reykjavik's city buses from internal combustion to fuel-cell engines, using hydroelectricity to electrolyze water and produce hydrogen. The next stage was to convert the country's automobiles, then its fishing fleet. These conversions were part of a systemic plan to divorce Iceland's economy from fossil fuels (Brown 2006, 168).

An industrial-scale hydrogen-fired power plant was being built near Venice during 2007 by the Veneto regional government and the Italian energy company ENEL. The new plant, in the Porto Marghera industrial area on the Italian mainland across from the Venice lagoon, next to ENEL's coal-fired Fusina plant, was designed as a "zero-emission hydrogen combustion power generation system" (Environment News Service 2006b).

Regardless of hydrogen fuel cells' limitations, the European Union has advocated a transition to them from fossil fuels. The plan includes a $2 billion EU commitment, over several years, to bring industry, the research community, and government together to solve technological problems. According to Rifkin:

> The EU decision to transform Europe into a hydrogen economy over the course of the next half century is likely to have as profound and far-reaching an impact on commerce and society as the changes that accompanied the harnessing of steam power and coal at the dawn of the industrial revolution and the introduction of the internal-combustion engine and the electrification of society in the 20th century. (*Industrial Environment* 2002)

## HYDROGEN POWER AND STRATOSPHERIC OZONE DEPLETION

Advocates of a hydrogen economy generally have ignored another potential problem. Some research indicates that leakage of hydrogen gas could cause problems in the Earth's stratospheric ozone layer. Writing in *Science*, Tracey K. Tromp and colleagues from the California Institute of Technology reported that the accumulation of leakage associated with a hydrogen economy could indirectly cause as much as a 10 percent decrease in stratospheric ozone.

> The widespread use of hydrogen fuel cells could have heretofore unknown environmental impacts due to unintended emissions of molecular hydrogen, including an increase in the abundance of water vapor in the stratosphere (plausibly as much as about 1 part per million by volume). This would cause stratospheric cooling, enhancement of the heterogeneous chemistry that destroys ozone, an increase in noctilucent clouds, and changes in tropospheric chemistry and atmosphere-biosphere interactions. (Tromp et al. 2003, 1740)

If hydrogen replaced all fossil fuels for transportation and to power buildings, Tromp and colleagues estimated that 60 to 120 trillion grams of hydrogen would be released into the atmosphere each year, four to eight times the amount released today from human sources. The scientists assumed a 10 to 20 percent loss rate due to leakage. Molecular hydrogen rises and mixes in the stratosphere, resulting in creation of moisture at high altitudes that through a chain of chemical reactions cools the air and accelerates destruction of ozone (Environment News Service 2003a). The researchers estimated that hydrogen leakage could triple the amount of hydrogen in the stratosphere.

"We have an unprecedented opportunity this time to understand what we are getting into before we even switch to the new technology," said Tromp, the study's lead author. "It will not be like the case with the internal combustion engine, when we started learning the effects of carbon dioxide decades later" (Environment News Service 2003a). Refuting assertions that hydrogen leakage could deplete ozone, Martin G. Schultz and colleagues, also writing in *Science*, asserted that "a possible rise in atmospheric hydrogen concentrations is unlikely to cause significant perturbations of the climate system" (Schultz et al. 2003, 624).

## REBIRTH OF THE ELECTRIC CAR

One of Henry Ford's fantasy cars was electric, but such designs were muscled aside by development of powerful internal combustion engines during the 1930s. Reacting to oil embargoes and gasoline price rises during the 1970s, the Ford Motor Company again developed workable designs for small, efficient, electric cars, and once again shelved them. The problem is less technological than a matter of driver preference and corporate profitability.

Electric cars may be coming back, however. In January 2007, General Motors rolled out the hybrid Chevrolet Volt, a concept car that can run 40 miles on electricity alone with a six-hour nighttime charge and, running as a hybrid, gets 150 miles per gallon of gasoline. For commuting trips less than 20 miles each way, it can run on a charge from a standard 110-volt garage outlet (Griscom-Little 2007, 60). Electricity must be generated, of course, and these days that usually involves the burning of fossil fuels.

By mid-2007, GM had committed to hiring four hundred technical experts to work on fuel-saving technology. One of the company's goals was to move the Volt to production within three or four years. This was quite a change from 2002, when General Motors introduced the gas-guzzling Hummer H2, which was so thirsty that the company took advantage of a loophole in federal law and refused to publish its mileage rating (Boudette 2007, A-8). At the new, green GM, Lawrence Burns, vice president for research, development, and global planning, told the *Wall Street Journal:* "We have to have people think we are part of the solution, not part of the problem." The Volt, said Burns, is an effort to show consumers that "we get it" on climate change (Boudette 2007, A-8).

With all the eco-hype over electric cars, a few things often have gone unsaid. One is that electricity has to come from somewhere. Today, most of it is generated by fossil fuels, half of it by coal, which is a dirtier greenhouse source per unit of energy than oil. If and when electric cars become *really* popular, they could create problems for electric utilities' peak-demand periods. Suppose large numbers of commuters plug in their cars at dinnertime on a hot evening when electricity demand is already high? The surge could overwhelm the utilities' capacity and lead to blackouts.

In addition, electric cars lack the driving range, interior room, and get-up-and-go performance of gas-powered automobiles. Costs, at $50,000 to $100,000, are prohibitive for all but showcase uses, although mass production would reduce sticker shock.

A safe, affordable 100,000-mile lithium-ion battery (less powerful versions are used in cell phones and laptops) is still in the future. California's Air Resources Board calculates that lithium-ion packs would cost $3,000 to $4,000 in mass production, cheap enough to be feasible (Ulrich 2007). The Chevy Volt's gasoline engine charges batteries, to a range of 640 miles. In our time, no purely electric vehicle with four seats and the ability to reach highway speeds has been mass-produced. Existing electric cars have been ridiculed as "glorified golf carts" (Ulrich 2007).

## NEW YORK CITY'S GREEN YELLOW CABS

New York City's taxi fleet will run entirely on gas-electric hybrids within five years, Mayor Michael Bloomberg announced May 22, 2007. "There's an awful lot of taxicabs on the streets of New York City," he said. "These cars just sit there in traffic sometimes, belching fumes" (Associated Press 2007f). Almost four hundred hybrids were tested in New York's taxi fleet over 18 months, with models including the Toyota Prius, Toyota Highlander Hybrid, Lexus RX 400h, and Ford Escape. Under Bloomberg's plan, that number will increase to 1,000 by October 2008, then will grow by about 20 percent each year until 2012, when every Yellow Cab—currently numbering 13,000—will be a hybrid (Associated Press 2007f). The city sells taxi licenses to individual drivers, who then purchase their own vehicles under specifications set by the Taxi and Limousine Commission. A similar hybridization of cabs was under way in San Francisco.

In 2006, the average New York taxi ran entirely on gasoline and got 14 mpg. By 2008, all New York taxis will be required to run at 25 mpg, and 30 mpg by 2009. While hybrid vehicles cost more, Bloomberg said that increased fuel efficiency will reduce operating costs by about $10,000 a year per vehicle. Changing the New York City taxi fleet to hybrids is one piece of Bloomberg's sustainability plan to reduce carbon dioxide emissions in the city 30 percent by 2030.

## CONGESTION CHARGES

Mayor Bloomberg also has proposed a congestion charge for the crowded (and often gridlocked) southern half of Manhattan Island, roughly from 86th Street southward. On weekdays from 6 A.M. to 6 P.M., trucks would be charged $21 a day and cars $8. The maximum congestion charge would affect only 5 percent of people who live outside Manhattan and work there who commute by car. People who drive solely within the zone would pay half price; taxis and livery cabs would be exempt.

With uncontrolled free access to the area, studies have shown that vehicle speeds within this area average 2.5 to 3.7 miles an hour. Many times, walking is almost as fast. Anyone who wants to go anywhere with any speed in this area takes the subway. The value of time lost to congestion delays in New York City has been estimated at $5 billion a year; add wasted fuel, lost revenue, and increasing costs of doing business, and the total rises to $13 billion a year (Kolbert 2007, 24). Despite these statistics, Bloomberg's plan was derailed by the New York State Assembly during July 2007.

In the meantime, other U.S. cities (among them San Francisco, Dallas, Miami, Minneapolis, and San Diego) were considering similar charges. London introduced a congestion charge in 2003. Singapore has one as well, as do Stockholm, Oslo, and Santiago, Chile.

In London, vehicle speeds increased 37 percent after the charge was implemented. London's mayor, Ken Livingstone, a major proponent of the charge, was easily reelected in 2004. By 2006, two-thirds of London residents supported the charge. During January 2007, the congestion zone was expanded westward to include most of Kensington, Chelsea, and Westminster (Kolbert 2007, 23–24).

By early 2007, London's congestion charge had reduced private car use 38 percent and carbon emissions 20 percent in the congestion charge zone. Motorcycles, taxis, police cars, bicycles, cars using alternative fuels, buses, and ambulances are exempt. Residents who live inside the tax zone get a 90 percent discount. International diplomats are exempt as well; the U.S. Embassy had caused a furor by racking up $3.4 million in unpaid fines by mid-2007 (Patrick 2007, W-4).

The London charge is not assessed at tollbooths but by cameras that ring the zone, reading license plates and checking them against computer records. Car owners may pay by the year; those who don't pay within a day of a citation for noncompliance (at convenience stores, by cell phone, or on the Internet) are fined £50 (about $100); the fine doubles after two weeks. The system cost $184 million to operate during 2006, but earned $430 million in fines (Patrick 2007, W-4).

London also improved its public transport system to offer Londoners easier alternatives. Many commuters had complained that the buses were slow and expensive; by the time the London bus fleet was upgraded, more than six million people were using it daily. The number of people commuting by bicycle soared 80 percent after the congestion charge was implemented (Environment News Service 2007c).

In the meantime, British Petroleum (whose corporate initials have been rendered by its publicists as "Beyond Petroleum") has proposed that every motorist in Great Britain sign up for a plan it calls "Target Neutral." Drivers can fund ventures that offset the amount of carbon dioxide that their driving adds to the atmosphere. Drivers register at a Target

Neutral website, which calculates the estimated amount of carbon dioxide that may be produced by their driving for a year. Drivers then pay offsets based on the estimate. The typical family car, driven 10,000 miles a year, is likely to cost about £20 ($40) to offset (Environment News Service 2006a).

Congestion charging in downtown Stockholm became a controversial issue in the Swedish general election during the late summer of 2006. A charge of up to $7 a day was narrowly approved by 52 percent in a referendum on September 17, 2006. The congestion charges reduced auto traffic 20 to 25 percent, while use of trains, buses, and Stockholm's extensive subway system increased. Emissions of carbon dioxide declined 10 to 14 percent in the inner city and 2 to 3 percent in Stockholm County. The project also increased the use of environmentally friendly cars (such as hybrids), which are exempt from congestion taxes. As in London, commuting by bicycle also increased. Following the trial period, the Stockholm congestion tax became permanent during August 2007.

Per Bolund, one of 19 Green Party members in Sweden's 349-member Riksdag parliament as of 2007, has been watching Green Party initiatives work their way into the political mainstream for many years. The Green Party favored a Stockholm congestion charge for decades, but conservatives blocked it. Ironically, the congestion charge was imposed under a conservative-oriented coalition government.

## RIDING THE RAILROAD: TRANSPORT OF THE FUTURE

Some of the hottest (which is to say, when it comes to preventing climate change, coolest) forms of transport for the twenty-first century are inventions of the nineteenth—the bicycle, that ultimate energy conservation machine, and the passenger railroad. Both have significant room to grow in the United States, where public transportation accounts for only 1 percent of total transport miles.

In the United States, by 2000, air travel had nearly replaced railroads for all but heavy freight. However, with mounting concern vis-à-vis carbon footprints and the decay of the U.S. commercial aviation system (prone to erratic schedules, reliance on expensive oil-based fuel, weather problems summer and winter, security paralysis, and other dysfunctions), punctual European trains look better every day. Passengers board trains within a few minutes of departure, while airline passengers are advised to arrive hours early to endure long lines at check-in, security, and boarding gates. European trains are clean, fast, spacious, and attended by courteous staff. Remodeled cars include trays for desktops, rooms for small meetings, and Wi-Fi access.

Some European trains also are breathtakingly fast, benefiting from a network of "dedicated track" 2,912 miles long that allows no freight or slower trains. France's TGV (Train à Grande Vitesse), which connects Paris with Germany's high-speed InterCityExpress, travels at 200 mph, with higher bursts. The French have turned speed trains into a spectator sport. In April 2007, an experimental TGV on the Paris–Strasbourg route hit 357 mph as cheering crowds lined the tracks. Meanwhile, Spain is building high-speed trains that will connect Madrid and Barcelona (375 miles, a distance that takes seven hours on Amtrak) in two and a half hours.

China has built a magnetic-levitation shuttle between the Pudong airport and downtown Shanghai that accelerates to 240 mph during the eight-minute trip. Plans call for a similar line to open between Beijing and Shanghai in 2010 (Finney 2007, 16); the trip, covering a distance equal to the 18-hour Amtrak ride from Chicago to Washington, D.C., will take five hours. China has set aside $250 billion to improve rail service. Japan has long used 180-mph "bullet trains" between Tokyo and Osaka, as well.

## TRAINS IN THE UNITED STATES

Having returned from Europe, my fond memories of train travel there dissolved into gritty, track-rattling reality in the United States. As with automobile mileage standards, when it comes to passenger trains, the rest of the world has taken off and left the United States behind—in the case of trains, *way* behind.

Outside the Northeast Corridor (Boston to Washington) passenger rail barely exists in the United States. The country has long had a skimpy national rail network (Amtrak), but trains are few, schedules are inconvenient, and tracks are often poorly maintained. The system is heavily tax subsidized, and Congress often threatens to put it out of business. In the biggest eastern cities (e.g., Boston, Washington, and New York), subways

**Figure 1.3. A magnetic levitation shuttle departing from the Pudong International Airport, Shanghai.**

Source: Wikipedia.com. Photo by Alex Needham.

and commuter trains are extensive and widely used. As in London, these run on electric rails (mind the gap!). In most other cities, subways do not exist. Nearly all intracity and long-distance travel is by car or airline.

Even in the United States, however, trains have been eating into airplanes' share of travel on some routes, notably between Boston, New York City, and Washington, D.C., where Amtrak ridership in July 2007 was up 20 percent over a year earlier, as airlines experienced scheduling problems and delays, not to mention concern over their carbon footprint. Passenger rail traffic between Chicago and St. Louis surged 53 percent during the same period (Machalaba 2007, B-1). On routes shorter than 400 miles, the airlines' time advantage has disappeared due to increasing security screenings, lines, and other problems. Amtrak since 2000 has operated Acela locomotives on the Northeast Corridor that can run up to 150 mph on parts of their routes. During 2007, the trains carried more passengers than airlines between Boston and Washington, D.C. Recognizing the increasing appeal of railroads for passenger traffic, the Senate late in 2007 authorized a six-year, $11.4 billion budget for Amtrak that increased its budget to nearly $2 billion a year, from $1.3 billion.

## BIKES IN THE NEW URBAN UTOPIA

Despite its health benefits (a regular bicycle rider's physiological age is 10 years younger than a person who usually drives a car), bicycle transport accounts for only 0.2 percent of travel miles in the United States, or one mile of every 500 (Hillman, Fawcett, and Rajan 2007, 53–54). At the same time, urban life in Europe is being recast with the automobile as antithesis. A driver is free to buy an SUV in Denmark, but the bill includes a registration tax of up to 180 percent of the purchase price. Imagine, for example, paying more than $80,000 in taxes (as well as $6 a gallon for gasoline as of 2007) to buy and drive a Hummer H2, only to watch pesky bicyclists ridicule his or her elegantly pimped ride as an environmental atrocity. Along similar lines, late in 2007 Britain's government was considering a graduated "purchase tax" that would reward fuel efficiency in automobiles, from a $4,000 penalty for the most inefficient vehicles to a $4,000 tax rebate for the best.

The automobile's urban territory has been shrinking in European cities. A growing web of pedestrian malls allows tens of thousands of people to traverse downtown Stockholm on foot every day—down a gentle hill, northwest to southeast, along Drottinggatan, past the Riksdag and the king's palace, merging with Vasterlanggatan, into the Old Town—for more than two miles. More and more streets across the city are gradually being placed off-limits to motor traffic (Johansen 2007, 23). Bicycles also account for one-eighth of urban travel in Sweden, where Stockholm is laced with many well-used bicycle paths that complement its growing web of pedestrian-only malls.

Bicycles have become privileged personal urban transport in many European cities. To sample bicycle gridlock, visit Copenhagen, which has deployed two thousand bikes around the city for free use. The mayor,

Klaus Bondam, commutes by bicycle. Helmets are not required, despite the occasional bout of two-wheeled road rage as bicyclists clip each other on crowded streets. People ride bikes, rain or shine, while pregnant, drinking coffee, or smoking, using a wide array of baskets to carry groceries and briefcases. On weekends, more than half the admissions to the emergency room of Frederiksberg Hospital are drunken cyclists, many of whom have rammed utility poles. On a more sober note, more than a third of Copenhagen residents ride bikes to work (40 percent do so in Amsterdam), in a conscious assault on the "car culture" (Keates 2007, A-10).

The Copenhagen airport has parking spaces for bicycles. New bike-parking facilities are planned at the Amsterdam's main train station that will house as many as ten thousand bikes. Officials from some U.S. cities, as well as some larger cities in Europe (London and Munich are examples) have been studying Amsterdam and Copenhagen.

Many Danish companies offer indoor bike parking, as well as locker rooms. Employees ride company-owned bikes to off-site meetings. People carry children on extra bike seats. Dutch prime minister Jan Peter Balkenende rides to work some of the time. Members of the Danish Parliament ride as well, along with CEOs of some major companies. Lars Rebien Sorensen, CEO of the pharmaceutical firm Novo Nordisk, conducts media interviews from his bike saddle.

More than 10,600 gray bicycles became available in Paris for modest rental prices under the Vélib program (for *vélo*, bicycle, and *liberté*, freedom) as of July 16, 2007, stationed at 750 self-service racks with instructions in eight languages. The number of bikes increased to 20,600 by the end of December 2007, at 1,450 stations, one for every 250 yards across the entire city. Users may rent a bike online or at any station, and return

**Figure 1.4. Vélib bicycles, part of a bicycle-sharing program, in Paris near a metro Cité (Station n° 4002 place Louis Lépine).**
Source: Wikipedia.com. Photo by Rcsmit. http://en.wikipedia.org/wiki/Image:Velibvelo1.jpg.

it at any other station, using a credit or debit card. Any ride of less than half an hour is free. A one-day pass costs €1 (about $1.50), a weekly pass €5 ($7.50), and a yearly subscription €29 ($43).

"This is about revolutionizing urban culture," said Pierre Aidenbaum, mayor of Paris's Third District. "For a long time cars were associated with freedom of movement and flexibility. What we want to show people is that in many ways bicycles fulfill this role much more today" (Bennhold 2007b). Vélib is part of a plan initiated by Socialist Paris mayor Bertrand Delanoë to reduce car traffic in the city by 40 percent by 2020.

Lyon, France's third largest city, launched a similar system during 2005. "It has completely transformed the landscape of Lyon—everywhere you see people on the bikes," said Jean-Louis Touraine, Lyon's deputy mayor. The program was meant "not just to modify the equilibrium between the modes of transportation and reduce air pollution, but also to modify the image of the city and to have a city where humans occupy a larger space." Jean-Luc Dumesnil, his Paris mayoral aide, added, "We think it could change Paris's image—make it quieter, less polluted, with a nicer atmosphere, a better way of life" (Anderson 2007, A-10).

The bikes were installed after a study analyzed travel in Paris by car, bicycle, taxi, and walking. Bikes were the quickest form of urban transport, as well as the least noisy and most environmentally friendly. The Lyon rental bikes, with their distinctive silver frames, red rear-wheel guards, handlebar baskets, and bells, also are among the least expensive modes of travel, because the first half-hour is free and most of 20,000 daily trips are shorter than that, according to Anthonin Darbon, director of Cyclocity, which operates Lyon's program. The same company also won the contract for the Paris program (Anderson 2007, A-10).

Cyclocity is a subsidiary of the outdoor advertising company JCDecaux, which operates much smaller bike-rental businesses in Brussels, Vienna, and the Spanish cities of Cordoba and Girona. London, Dublin, Sydney, and Melbourne are considering similar programs (Anderson 2007, A-10). These systems evolved from utopian "bike-sharing" ideas that were tried in Europe in the 1960s and 1970s, patterned on Amsterdam's "white bicycle" plan, in which volunteers repaired hundreds of broken bicycles, painted them white, and left them on the streets for free use. Many of these bikes were stolen, and others broke down from lack of maintenance. To avoid the same problems, writes John Ward Anderson described in the *Washington Post*:

> JCDecaux experimented with designs and developed a sturdier, less vandal-prone bike, along with a rental system to discourage theft: Each rider must leave a credit card [number] or refundable deposit of about $195, along with personal information. In Lyon, about 10 percent of the bikes are stolen each year, but many are later recovered, Darbon said. To encourage people to return bikes quickly, rental rates rise the longer the bikes are out. (Anderson 2007, A-10)

A number of U.S. cities, including Portland, Oregon, also have experimented with community-use bicycle programs (Anderson 2007, A-10). By late 2007, Chicago and San Francisco were setting up bike-sharing programs that emulate Europe's. New York City was making room on some of its roads for bike-only traffic lanes.

Portland, Oregon, which has been building bicycle lanes since the 1970s, has promoted itself as "Bike City, USA." It has a substantial manufacturing base, with about 125 bicycle-related companies. These include national companies such as Nike and Columbia Sportswear, which are headquartered there; Team Estrogen, which sells cycling clothing for women; and smaller businesses that build bike frames and other gear, including custom-made bicycles. A local a consulting firm, Alta Planning and Design, advises other cities on bicycle-friendliness (Yardley 2007).

Some U.S. cities are increasingly bike friendly. New York City's mayor Michael Bloomberg has proposed requiring commercial buildings to maintain indoor bike parking. Boulder, Colorado, has bike lanes on 97 percent of its arterials; one-fifth of people who commute there do so by bike. Boulder has been spending 15 percent of its transportation budget on bike infrastructure. The university town of Davis, California, which now has a bicycle on its city logo, as of 2007 had 95 percent arterials with bike lanes and a 17 percent two-wheeled commuter rate.

## REFERENCES

Anderson, John Ward. 2007. "Paris Embraces Plan to Become City of Bikes." *Washington Post*, March 24, A-10. http://www.washingtonpost.com/wp-dyn/content/article/2007/03/23/AR2007032301753_pf.html.

Andrews, Edmund L. 2007b. "Senate Adopts an Energy Bill Raising Mileage for Cars." *New York Times*, June 22. http://www.nytimes.com/2007/06/22/us/22energy.html.

Associated Press. 2007f. "New York's Mayor Plans Hybrid Taxi Fleet." *New York Times*, May 22. http://www.nytimes.com/aponline/us/AP-Green-Taxis.html.

Bennhold, Katrin. 2007b. "Paris Journal: A New French Revolution's Creed: Let Them Ride Bikes." *New York Times*, July 16. http://www.nytimes.com/2007/07/16/world/europe/16paris.html.

Bjerklie, David. 2007. "Check Your Tires." *Time*, April 9. http://www.time.com/time/printout/0,8816,1603740,00.html.

Boudette, Neal E. 2007. "Shifting Gears, General Motors Now Sees Green." *Wall Street Journal*, May 29, A-8.

Brown, Lester R. 2006. *Plan B: Rescuing a Planet under Stress and a Civilization in Trouble*, rev. ed. New York: Earth Policy Institute/W. W. Norton.

Bunkley, Nick. 2007a. "Detroit Finds Agreement on the Need to Be Green." *New York Times*, June 1. http://www.nytimes.com/2007/06/01/business/01auto.html.

———. 2007b. "Seeking a Car That Gets 100 Miles a Gallon." *New York Times*, April 2. http://www.nytimes.com/2007/04/02/business/02xprize.html.

Chandler, Michael Alison. 2007. "Without a Car, Suburbanites Tread in Peril: Loudoun Residents Blaze Their Own Risky Trails Where Sidewalks and Bike Paths Are Lacking." *Washington Post*, July 16, B-1. http://www.washingtonpost.com/wp-dyn/content/article/2007/07/15/AR2007071501345_pf.html.

Cline, William R. 1992. *The Economics of Global Warming*. Washington, D.C.: Institute for International Economics.

Commoner, Barry. 1990. *Making Peace with the Planet.* New York: Pantheon.

Coy, Peter. 2002. "The Hydrogen Balm? Author Jeremy Rifkin Sees a Better, Post-petroleum World." *Business Week*, September 30, 83.

Environment News Service. 2003a. "Hydrogen Leakage Could Expand Ozone Depletion." June 13. http://ens-news.com/ens/jun2003/2003-06-13-09.asp.

———. 2006a. "British Travel Agents Launch Carbon Offset Scheme." November 28. http://www.ens-newswire.com/ens/nov2006/2006-11-28-05.asp.

———. 2006b. "Italy to Build World's First Hydrogen-Fired Power Plant." December 18. http://www.ens-newswire.com/ens/dec2006/2006-12-18-05.asp.

———. 2007b. "Automakers Join Call for National Greenhouse-Gas Limits." June 27. http://www.ens-newswire.com/ens/jun2007/2007-06-27-09. asp#anchor4.

———. 2007c. "Big City Mayors Strategize to Beat Global Warming." May 15. http://www.ens-newswire.com/ens/may2007/2007-05-15-01.asp.

———. 2007h. "California Air Board Adds Climate Labels to New Cars." June 25. http://www.ens-newswire.com/ens/jun2007/2007-06-25-09.asp#anchor7.

———. 2007w. "Mandatory U.S. Greenhouse Gas Cap Wins New Corporate Supporters." May 8. http://www.ens-newswire.com/ens/may2007/2007-05-08-01.asp.

———. 2007cc. "Volvo First Automaker to Go Carbon Dioxide Free." September 24. http://www.ens-newswire.com/ens/sep2007/2007-09-24-03.asp.

Finney, Paul Burnham. 2007. "U.S. Business Travelers Let the Train Take Way the Strain." *International Herald-Tribune*, April 24, 16.

Gordon, Anita, and David Suzuki. 1991. *It's a Matter of Survival.* Cambridge: Harvard University Press.

Grant, Paul M. 2003. "Hydrogen Lifts Off—with a Heavy Load: The Dream of Clean, Usable Energy Needs to Reflect Practical Reality." *Nature* 424 (July 10): 129–30.

Griscom-Little, Amanda. 2007. "Detroit Takes Charge." *Outside*, April, 60.

Hakim, Danny. 2004. "Several States Likely to Follow California on Car Emissions," *New York Times*, June 11, C-4.

Hillman, Mayer, Tina Fawcett, and Sudhir Chella Rajan. 2007. *The Suicidal Planet: How to Prevent Global Climate Catastrophe.* New York: St. Martin's/Dunne.

Hitt, Greg. 2007. "Changed Climate on Warming." *Wall Street Journal*, March 20, A-6.

*Houston Chronicle.* 2007. "Hydrogen Car Has Far to Go." Reprinted in *Omaha World-Herald*, September 9, 2-D.

*Industrial Environment.* 2002. "E.U. Plans to Become First Hydrogen Economy Superpower." 12, no. 13 (December), n.p. (in LEXIS).

Johansen, Bruce. 2007. "Scandinavia Gets Serious about Global Warming." *Progressive*, July, 22–24.

Keates, Nancy. 2007. "Building a Better Bike Lane." *Wall Street Journal*, May 4, W-1, W-10.

Kolbert, Elizabeth. 2007. "Don't Drive, He Said." Talk of the Town. *New Yorker*, May 7, 23–24.

Lippert, John. 2002. "General Motors Chief Weighs Future of Fuel Cells." *Toronto Star*, September 27, E-3.

Lovins, Amory. 2005. "More Profit with Less Carbon." *Scientific American*, September, 74, 76–83.

Machalaba, Daniel. 2007. "Crowds Heeds Amtrak's 'All Aboard.'" *Wall Street Journal*, August 23, B-1, B-22.

Masters, Coco. 2007b. "Fill'er Up with Passengers." *Time*, March 27. http://www.time.com/time/printout/0,8816,1603736,00.html.

McKibben, Bill. 1989. *The End of Nature*. New York: Random House.

Patrick, Aaron O. 2007. "Life in the Faster Lane." *Wall Street Journal*, July 20, W-1, W-4.

Romm, Joseph J. 2007. *Hell and High Water: The Solution and the Politics—and What We Should Do*. New York: William Morrow.

Sayre, Caroline. 2007. "Make One Right Turn after Another." *Time*, April 9. http://www.time.com/time/printout/0,8816,1603741,00.html.

Schultz, Martin G., Thomas Diehl, Guy P. Brasseur, and Werner Zittel. 2003. "Air Pollution and Climate-Forcing Impacts of a Global Hydrogen Economy." *Science* 302 (October 24): 624–27.

Spector, Mike. 2007. "Can U.S. Adopt Europe's Fuel-Efficient Cars?" *Wall Street Journal*, June 26, B-1.

Steinman, David. 2007. *Safe Trip to Eden: 10 Steps to Save Planet Earth from Global Warming Meltdown*. New York: Thunder's Mouth Press.

Surowieki, James. 2007. "Fuel for Thought." *New Yorker*, July 23, 25.

Tromp, Tracey K., Run-Lie Shia, Mark Allen, John M. Eiler, and Y. L. Yung. 2003. "Potential Environmental Impact of a Hydrogen Economy on the Stratosphere." *Science* 300 (June 13): 1740–42.

Ulrich, Lawrence. 2007. "They're Electric, but Can They Be Fantastic?" *New York Times*, September 23. http://www.nytimes.com/2007/09/23/automobiles/23AUTO.html.

White, Joseph R. 2006. "An Ecotopian View of Fuel Economy." *Wall Street Journal*, June 26, D-4.

Yardley, William. 2007. "In Portland, Cultivating a Culture of Two Wheels." *New York Times*, November 5. http://www.nytimes.com/2007/11/05/us/05bike.html.

CHAPTER 2

# Avoiding Our "Suicide Pact" with Aviation

I once met Ken Lincoln, an English professor at the University of California at Los Angeles who lives in Santa Fe and commutes—inbound by air on Mondays for a Tuesday–Thursday schedule—outbound on Fridays, perhaps three dozen times a year. This was the first evidence I had seen of aviation-aided urban sprawl. The eastern suburbs of Los Angeles had spread across parts of three states, nearly 800 miles, into the heart of New Mexico. Lincoln said it beat an hours-long daily grind past pricey Beverly Hills and the gilded Hollywood Hills to Burbank, the only decent housing a UCLA professor could afford.

Lincoln was not alone. During the 1990s, a rising number of people in the United States were commuting to work via the airlines, sometimes thousands of miles per week. The *New York Times* carried accounts of Manhattan jobholders flying into the city from Rochester, New York. Silicon Valley, where the price of the average house had broached $600,000 by 2006, was drawing weekly commuters, according to one *New York Times* account, from "Arizona, Idaho, Nevada, Oregon, and Utah" (Johnston 2000, 1-G). Some people who work at New York's Lincoln Center commute from Florida. The typical schedule includes long workdays Tuesday through Thursday, travel Monday and Friday, and a weekend at home hundreds, sometimes thousands, of miles from the office.

For those with a penchant for technology and no concern about rising greenhouse-gas levels in the atmosphere, there's more. Casting aside air travel's effects on the atmosphere, I read in some newspapers' "style" sections that the status vehicle in elite circles of the future won't be a sports-utility vehicle. According to one observer, it will be "the family plane" (Fallows 1999, 84). Just such an aircraft, the Cirrus SR20, was being produced by the late 1990s, as the National Aeronautics and Space

Administration (NASA) was "quietly advocating an Interstate Skyway Network" (Fallows 1999, 88). Between 2000 and 2005, the number of private jets in the United States grew by 40 percent. By 2007, ten thousand private jets already were registered in the United States, each burning as much as fifteen times as much fuel per passenger-mile as commercial flights (Frank 2007, W-2).

I read features in the *Wall Street Journal* promoting "flying cars"—the Jetsons come to life—with no mention of their impact on the atmosphere. Environmentally, "air cars" would make Hummers look like bicycles. The flying car to some people is the next step beyond the private jet.

By 2006, about a hundred people had paid $25,000 each to reserve a Moller International "Skycar," which was at the test stage. Skycars are expected to have an initial sticker price of about $500,000 each (Stoll 2006, R-8). The Skycar is designed to taxi from home to a designated takeoff area (a "vertiport") laid out on a parking lot or field. The Skycar, which its builders say will sell for about $60,000 once it is in mass production, will run on methanol, ethanol, diesel, or gasoline.

## THE CARBON FOOTPRINT OF AVIATION

I had no heart to calculate Professor Lincoln's carbon footprint. Besides, I should watch out when I throw rocks at glass houses—or airport terminals. While I am a climatic good boy most days on my four-mile round-trip commute from home to the University of Nebraska at Omaha's campus on my mountain bike, now and again I find myself accepting invitations to lecture in Wales and Poland or going on a magazine assignment (on combating global warming, no less) in Sweden. On the butt-squaring journey across the Atlantic, I have been responsible for as much greenhouse-gas emission as I would have produced in an ordinary car commuting to and from my campus from the western suburbs of Omaha in an entire year. I am, thus, contributing my own share of frequent-flyer miles to Americans' enormous and growing carbon overload from aviation. Air mileage per person in the United States increased 400 percent between 1970 and 2006 (Hillman, Fawcett, and Rajan 2007, 55). The rest of the world is not far behind.

During the spring of 2006, I found myself at the University of Swansea in Wales, lecturing on cultural attributes of Native American humor. On the train back to London, I stopped in Oxford to visit Mark Lynas, author and climate activist, who had just published a cover story in the *New Statesman* titled "Our Suicide Pact with Aviation." He had two copies and gave me one, which I read amid surging crowds on a Sunday morning at Gatwick, one of London's three burgeoning international airports, where a new terminal was being constructed to handle increasing traffic.

A few days earlier, a large headline on the front page of London's *Independent*, the most aggressive of Britain's newspapers on global warming, had upbraided Prince Charles for talking up the risks of global warming as he flew to India on a private jet, an Airbus 319. The *Independent* described, in relishing detail, a round-trip by the Prince from London to India via Egypt and Saudi Arabia that covered 9,272 miles, emitting 42 tons of carbon dioxide (Hickman 2006, 1). At the same time, Charles was arguing that global warming was "the greatest challenge" facing humankind. The newspaper urged him to walk the walk.

By 2007, Prince Charles disclosed that his personal carbon footprint had shrunk 9 percent in a year to 3,775 tons. He had been taking private jets fewer times (substituting trains when available) and gassing up the royal Jaguar with cooking oil. Otherwise, Charles has gone "carbon neutral" with offsets worth $600,000 a year. Still, his three mansions had a carbon footprint the size of five hundred average British homes (*Wall Street Journal* 2007, A-14).

As I read Lynas's article at Gatwick, flight traffic was exploding in England, with a third runway planned at London's Heathrow and similar extensions at London's Stansted, as well as Birmingham, Edinburgh, and Glasgow. Twelve other British airports also had announced expansion plans. According to the House of Commons' Environmental Audit Committee, increasing airline passenger traffic in Great Britain would require the equivalent of another Heathrow-size airport every five years (Monbiot 2006b). Heathrow itself was planning a sixth terminal while its fifth is still was under construction. The crowds at Gatwick, even early on a Sunday morning, reminded me of Manhattan streets outside Macy's on Christmas Eve—sheer human gridlock.

Budget airlines such as Ryanair and EasyJet contributed to a 76 percent increase in traffic through Britain's airports in a decade, to 215 million passengers in 2004 (Clark 2006). Even as they talked about reining in greenhouse-gas emissions, political leaders in the United Kingdom were promoting rapid increases in air travel and expansion of airports, much as automobile manufacturers in the United States talked up greenhouse-gas emissions caps while opposing higher gas-mileage standards. Both were "green" in theory and anything but in practice. As Prime Minister Tony Blair was saying, "Climate change is, without doubt, the major long-term threat facing our planet," British air mileage continued to explode (Lynas 2006, 12).

As I read Lynas's challenge to the British aviation juggernaut, I was sitting in the midst of the world's largest air travel hub. One-fifth of the world's international airline passengers fly to or from an airport in the United Kingdom. The number of passengers in this hub had risen fivefold in the thirty years ending in 2005, and the government forecast that they will more than double again by 2030, to 476 million a year (Monbiot 2006b). Carbon dioxide emissions from air traffic originating in the United Kingdom rose 85 percent between 1990 and 2000, according to its Department of Trade and Industry (Houlder 2002, 2). According to a report published by the Institute of Public Policy Research, "Flying by jet plane is the least environmentally sustainable way to travel and transport goods" (Lean 2001, 23).

Air travel soared a hundredfold during the last half of the twentieth century and is expected to rise substantially during the first half of the twenty-first. Between 1969 and 1989, according to the International Energy Agency, worldwide airline passenger miles increased 400 percent (International Energy Agency 1993, 44). From 1990 to 2000, the number of miles flown by airline passengers doubled worldwide from 67.5 billion to 140 billion. Between April 2005 and April 2006, passenger traffic rose 9.9 percent. Given present growth trends, global air traffic may increase sixfold by 2050 (*Wall Street Journal* 2006, A-7).

Airline travel contributed an estimated 5 percent to the total global human-generated greenhouse gases in 1990, an amount that was increasing at a much faster rate than overall fossil-fuel usage. If the rate of world air travel increase is maintained, air travel may account for 15 percent of greenhouse-gas emissions by 2050 (Clover 1999).

The United States is the origin point for a third of the world's commercial aviation. By 2004, air travel in the United States was consuming 10 percent of all fossil-fuel energy, with passenger loads expected to double between 1997 and 2017, making air travel the fastest growing source of carbon dioxide and nitrous oxides in the economy (Flannery 2005, 282).

During the 1990s, aviation was the fastest-growing mode of travel in the United States. To handle this growth, thirty-two of the fifty busiest U.S. airports had plans to expand as of 1999. Sixty of the hundred largest airports were proposing to build new runways. During the late 1990s, passenger and freight air transport mileage was doubling roughly every ten years. By the late 1990s, more than 2 billion passengers and 42 million tons of cargo were being transported by air worldwide per year. The number of passengers is increasing by 8 percent a year, on average, as the volume of cargo rises by 13 percent annually.

Sitting in the human swirl at Gatwick, I read Mark Lynas's words in the *New Statesman*:

> It has often been said that unlimited growth is the ideology of the cancer cell. If so, then aviation's tumours are metastasising all over Britain. No major city today is complete without its own local airport, offering cheap flights to an ever-increasing list of domestic and international destinations.... As an unavoidable consequence, aviation emissions will double by 2020 and quadruple by 2050, a prospect that makes a mockery of all other national efforts to combat global warming. (2006, 12).

Lynas wrote that, according to a report by the Tyndall Centre for Climate Change Research,

> even if we were to shut down the rest of the economy in order to save on greenhouse-gas emissions, aviation alone would bust the sustainable emissions budget by the middle of the century. Without heating, lights, cars, factories or any of the other sources of pollution, the growth in flying alone will propel us into a future of melting ice caps, spreading deserts, rising sea levels, vanishing farmland and collapsing ecosystems. (2006, 12)

"Just look at where the big money is being spent," he continued. "The private sector's price tag for Stansted's proposed runway is £2.7 billion, somewhere between ten and a hundred times the amount the government puts into its entire climate-change program, windmills, loft-insulation schemes and cycle lanes included" (2006, 14).

## AIR TRAVEL'S EFFECT ON THE ATMOSPHERE

Aircraft emissions are especially damaging because much of their pollution takes place in the upper atmosphere, at jet stream level. "As far as climate change is concerned," wrote George Monbiot in the London *Guardian*,

> this is an utter, unparalleled disaster. It's not just that aviation represents the world's fastest-growing source of carbon-dioxide emissions. The burning of aircraft fuel has a "radiative forcing ratio" of around 2.7. What this means is that the total warming effect of aircraft emissions is 2.7 times as great as the effect of the carbon dioxide alone. The water vapor they produce forms ice crystals in the upper troposphere (vapor trails and cirrus clouds) which trap the earth's heat. According to calculations by the Tyndall Centre for Climate Change Research, if you added the two effects together (it urges some caution as they are not directly comparable), aviation's emissions alone would exceed the government's target for the country's entire output of greenhouse gases in 2050 by around 134 percent. (Monbiot 2006b)

## ALTERING THE CHEMISTRY OF THE STRATOSPHERE

Air transport emits a toxic cocktail of gases. Atmospheric emissions from aircraft have almost three times the global-warming potential of their carbon dioxide content alone. In addition to $CO_2$, combustion of jet fuel injects water vapor and sulfur dioxide into the stratosphere, both of which enhance ozone depletion at that level. Nitrous oxides enhance ozone at lower levels, where it is a pollutant, while depleting it in the stratosphere, where it helps guard against ultraviolet radiation. The chemicals emitted into the atmosphere by burning aviation fuel "cause the formation of polar stratospheric clouds, affect markedly the aerosol composition of the atmosphere, and intensify the greenhouse effect" (Kondratyev, Krapivin, and Varotsos 2004, 249).

Projections that air traffic will double within two decades have compelled atmospheric scientists to ask whether burgeoning air traffic is altering the chemistry of the stratosphere through which jets travel. The combustion of jet fuel releases into the atmosphere several chemicals that affect the balance of greenhouse gases: carbon dioxide, water vapor, nitrous oxides, sulfur oxides, and particulate matter, notably soot (Friedl 1999, 57).

By 1999, in some air traffic corridors above Europe, jet contrails sometimes covered as much as 4 percent of the sky at any given time. In the northeastern United States, contrails sometimes covered as much as 6 percent. Contrail coverage over Asia was increasing so quickly that contrails could multiply by a factor of ten in fifty or sixty years. Jet contrails have a

**Figure 2.1. Jet contrails. © Lowell Sannes.**
Courtesy of Shutterstock.

net atmospheric warming effect similar to that of high thin ice clouds, trapping outgoing long-wave radiation. The contrails also reflect some incoming solar radiation. Given this balance, nighttime flights in winter (December though February) at a site in southeastern England "were responsible for most of the contrail radiative forcing" (Stuber et al. 2006, 864). Night flights "account for only 25 percent of daily air traffic, but contribute 60 to 80 percent of the contrail forcing" (Stuber et al. 2006, 864). Winter flights account for only 22 percent of annual air traffic, but contribute half of the annual mean forcing. "These results suggest that flight rescheduling could help to minimize the climate impact of aviation" (Stuber et al. 2006, 864).

## FLYING ABOVE THE LAW

Because airline travel often crosses international borders, its greenhouse-gas emissions have been excluded from many tallies on a nation-by-nation basis. The airline industry routinely uses its international status to avoid pesky national constraints such as energy-efficiency targets that go along with plans to reduce greenhouse-gas emissions. The 1944 Chicago Convention, now supported by four thousand bilateral treaties, rules that no government may levy tax on aviation fuel, giving the industry a unique incentive to avoid researching and enforcing efficiency where our transportation system needs it most. "The airlines," writes British author Monbiot, "have been bottle-fed throughout their lives" (2006b).

Jet fuel is one of the very few untaxed fuels in the world; it is also zero Value Added Tax (VAT) rated. The airline industry argues that it *is* effectively taxed, as British airlines pay a duty of £5 per passenger for a European flight (this fee was halved from £10 per flight in 2000), which raises £900 million (about $1.8 billion) per year. However, airlines also benefit from a £9 billion tax subsidy per year in the United Kingdom (Francis 2006).

Despite its growing role in global warming, aviation is unregulated and unrecognized by international climate treaties such as Kyoto, even as it grows faster than any other form of transport worldwide, averaging 12 percent a year. Air-pollution emissions from international flights also have been routinely excluded from treaties agreed to counter global warming and ozone depletion. Airports in Britain are exempt from pollution control and—in most cases—from statutory noise regulation. And, to top it all off, air travel receives billions of dollars every year in subsidies from taxpayers of several industrialized countries (Lean 2001, 23).

## AIR FUEL EFFICIENCY: JUST AN INKLING OF WHAT NEEDS TO BE DONE

Improvements in airline fuel efficiency offer some promise of incremental energy savings and pollution reduction. Flying 6,000 feet lower may save 6 percent of fuel, for example. Better planning to avoid long waits on takeoff and more direct routing may save 10 percent. Boeing has pitched its 787 Dreamliner as a marvel of efficiency that will stretch fuel supplies by 20 percent because it uses light carbon-fiber composites in place of aluminum. Even with some incremental gains in efficiency, however, today's mode of air travel remains a problem in any global-warming calculus. Every passenger on a long-haul flight is responsible for 124 kilograms (about 270 pounds) of carbon dioxide per hour (McGuire 2005, 188–89).

The air travel industry has reduced the amount of fuel it burns to transport each of its passengers by half since the mid-1970s, but the growth of air travel has more than canceled these savings in fuel efficiency. The same relationship applies to nitrogen oxides and hydrocarbons. According to the International Air Transport Association, jet engines by 2006 were 40 percent more fuel-efficient than they were in the 1960s. In 2006, U.S. airlines used about a billion fewer gallons of fuel than in 2000, yet carried 12 percent more passengers. American Airlines and Delta jets, for example, now sometimes taxi using only one engine to reduce fuel use, and these two carriers have promoted their efforts to modify wings so that they reduce drag and boost efficiency, as well as finding ways to reduce weight of aircraft by removing ovens, galleys, and water (Wilber 2007, D-1).

Will pigs fly? In the world of biomass fuel, they might. By 2007, the U.S. Department of Defense (DOD) and NASA were funding exploratory projects into biofuels for jet airplanes. Syntroleum was providing the DOD with jet fuel derived from animal fats supplied by Tyson Foods. Tyson is the world's largest producer of chicken, beef, and pork, producing prodigious amounts of animal fats such as beef tallow, pork lard, chicken fat, and greases—all of which may someday be used as fuel. The two companies have planned a plant at a thus-far undesignated location in the Southwest that after 2010 will produce 75 million gallons of fuel per year. According to Syntroleum, the U.S. Air Force plans to certify all its aircraft to run on alternative fuels by 2010 and wants 50 percent of its fuel to come from domestic alternative sources by 2016 (Environment News Service 2007p).

There even has been some talk of manufacturing jet fuel from soya (a kind of airline ethanol), which may reduce greenhouse-gas emissions slightly. I wonder how many square miles of soybean fields would be required to get a Dreamliner from New York City to London and back? There even has been some speculation that someday airliners may run on hydrogen fuel, after some pesky technological hurdles involving weight and thrust have been surmounted. For the time being, the hydrogen-powered aircraft remains in the realm of when-pigs-fly aviation.

## ATTEMPTS TO REDUCE AIRCRAFT EMISSIONS

The long life of jet aircraft blunts attempts to raise an airline fleet's overall fuel efficiency. The Boeing 747, for example, is still flying thirty-six years after it was introduced. The Tyndall Centre predicts that the Airbus A380, new in 2006, will be flying (in slightly modified form) in 2070. "Switching to more efficient models," wrote Monbiot, "would mean scrapping the existing fleet" (2006b). Future efficiency gains also may be meager due to the mature nature of jet-engine technology. Various systems have been tried and abandoned that might make the aerodynamics of jet aircraft more efficient. Usually, systems meant to improve laminar flow do not pay for themselves over the life of an aircraft.

At present, the use of hydrogen fuel or ethanol in place of the usual kerosene jet fuel faces formidable problems. Hydrogen fuel provides only 25 percent as much energy per volume as jet fuel, meaning that a hydrogen-powered aircraft would need huge fuel tanks and would have to fly with a heavier load, reducing mileage. Because of the volume, most fuel would be carried not in the wings but in the body of the aircraft, increasing drag. Hydrogen would produce no carbon dioxide, but its output of water vapor at high altitudes would increase the size of contrails, which aggravate global warming. Richard Branson, owner of Virgin Atlantic Airlines, during September 2006 revealed plans to invest $3 billion to develop ecologically friendly plant-based jet fuel. However, plant-based fuel weighs two-thirds more by volume than kerosene for the same amount of thrust. It also freezes easily at high altitudes (Daviss 2007, 35). On February 24, 2008, Virgin Atlantic flew a Boeing 747-400 from London to Amsterdam, partly on biofuel made from babassu nuts and coconut oil (mixed with standard jet kerosene), a first for a commercial aircraft.

The European Commission on December 20, 2006, proposed legislation to bring greenhouse-gas emissions from civil aviation into the EU Emissions Trading Scheme. The proposed law will cover emissions from flights within the European Union from 2011 and all flights to and from EU airports beginning in 2012.

The Association of British Travel Agents (ABTA) in late 2006 joined with two other British travel industry organizations in a carbon-offset program. The program allows agents to offer customers an opportunity to compensate for the climate-warming impact of their travel (notably airline trips) by contributing financially toward environmental projects around the world. ABTA joined with the Association of Independent Tour Operators and the Federation of Tour Operators to develop a plan that was to begin early in 2007. British Airways meanwhile has introduced a "$CO_2$ Emissions Calculator" on its website that allows passengers to pay an offset for their flights (Michael and Carey 2006, A-6).

Carbon offsets may help to pay for wind farms, development of more energy-efficient technology, new plantings of trees, or other projects. The concept's critics contend that greenhouse-gas emissions should be reduced in reality rather than offset in theory. Offsets allow guilt-free carbon dioxide indulgences (such as flying long distances) and parading of green corporate credentials, they argue. Writer Monbiot has said the most destructive effect of the carbon offset trade is that it allows people to believe they can continue polluting with complacency by paying a tree-planting company somewhere (Environment News Service 2006a).

An Intergovernmental Panel on Climate Change (IPCC) report on air travel and global warming stimulated calls for environmental taxes and stricter emissions targets for airlines. The report was produced jointly by the United Nations and the World Meteorological Organization. "Transportation is the area of greatest growth (in terms of impact on the climate), and aviation is growing more rapidly than any other sector," said John Houghton of Britain, one of the report's coauthors (Reuters 1999). "Policy options include more stringent aircraft engine emissions regulations, removal of subsidies and incentives [and] market-based options such as environmental levies," the report said (Reuters 1999).

## DAMAGE FROM "ECOTOURISM"

Steve McCrea, editor of the *Eco-Tourist Journal*, calls air travel "ecotourism's hidden pollution" (1996). Tourists who take the utmost ecological care when they visit exotic locales rarely give a second thought to the greenhouse gases that they generate while reaching their destinations. According to McCrea:

> One ton of carbon dioxide enters the atmosphere for every 4,000 miles that the typical eco-tourist flies. A round trip from New York to San Jose, Costa Rica (the world's leading eco-tourist destination) is 4,200 miles, so the typical eco-tourist generates roughly 2,100 pounds of carbon dioxide by traveling to a week of sleeping in the rainforest. (1996)

To balance the carbon dioxide generated by their air travel, McCrea suggests that ecotourists plant three trees for every 4,000 miles flown, to compensate not only for the carbon dioxide but also for other greenhouse gases created by the combustion of jet fuel. Given the damage to the atmosphere from aviation, however, three trees per 4,000 miles seems hardly even a symbolic remedy.

Environmental authorities in some countries are taking a closer look at the carbon footprint of aviation-based "eco-tours." In one case, Inkaterra advertised helicopter eco-tours from Cuzco, Peru, to Machu Picchu, which many visitors reach only after long hikes. Environmental authorities in Peru took a look at the carbon footprint of this "eco-tour" and, after a few flights during May 2007, shut it down (Higgins 2006, 6).

The same goes for long-distance tourism generally. "So as we are flying into the Alps for our ski holiday we are contributing to their destruction," wrote one European author. "Our honeymoon flight to the Maldives is slowly sinking it under rising sea levels and destroying coral through bleaching associated with global warming; and finally our safari flight to Africa is contributing to drought, famine and disease. It's not an appetizing thought, is it?" (Francis 2006).

Even as the glaciers melt, commission-conscious travel agents under the banner of ecotourism advise clients to hurry up and see them before they're gone. Betchart Expeditions of Cupertino, California, in 2007 offered a twelve-day tour to "Warming Island," off Greenland, which had recently emerged from melting ice, "a compelling indicator of the rapid speed of global warming" (Naik 2007a, A-12). The cost of the full tour was $5,000 to $7,000, plus airfare. In the Greenland coastal village of Illuissat, population 5,000, about 35,000 tourists arrived during 2007, most of them on cruise ships, up from 10,000 about five years previously. Many came to witness the receding Jacobshavin Glacier, which had lost nine miles in five years. Once upon a time, winter temperatures routinely fell to −40°F there; by 2007, −15°F was the usual winter's low. The harbor, which used to freeze solid, now remains liquid all year, allowing fisherman to pull halibut out of the water at all seasons, depleting stocks (Naik 2007a, A-12).

## A "NO-FLYING MOVEMENT"

So what can fliers who are concerned about ruining the atmosphere do? The elegantly simple—and perhaps, for the time being, only—effective answer is: don't fly as much. While airline manufacturers have gained some efficiency in recent decades, no nonfossil fuel provides the thrust that airliners need to gain and maintain altitude and speed. No one is seriously entertaining the idea of solar- or wind- (or, God help us, nuclear-) driven aircraft. Like no other form of transport, the modern aircraft is a hostage of the fossil-fuel age and a major producer of greenhouse gases.

"In researching my book about how we might achieve a 90 percent cut in carbon emissions by 2030," wrote Monbiot, "I have been discovering, greatly to my surprise, that every other source of global warming can be reduced or replaced to that degree without a serious reduction in our freedoms. But there is no means of sustaining long-distance, high-speed [air] travel" (2006b).

**Figure 2.2. View of south side, Terminal 5, London's Heathrow International Airport, which opened March 27, 2008.**
Source: Wikipedia.com. http://en.wikipedia.org/w/index.php?title=Image%3AZEMA0003.jpg.

In Britain, a "no-flying movement" has taken shape, as many people avoid aviation for all except essential trips. Newspaper travel supplements provide information on train and shipping alternatives. Vacationers were advised: "Travel to the Alps by train and you get a real sense of geography, of evolving culture and changing climatic zones. Arrive by air and all you see is identical airport terminals and thousands of other culture-shocked, aggravated travelers. Slow travel, like slow food, is about clawing back quality of life" (Lynas 2006, 14–15).

Environmental protesters in the United Kingdom have tried to halt expansion at several British airports, including Heathrow and Stansted. On February 19, 2006, a convoy of more than a hundred cars toured some of the villages to be affected by proposals to build a second runway at Stansted (Clark 2006). In 2007, hundreds of protesters at Heathrow camped out for several days in a "camp for climate action" and were dispersed by riot police. In February 2008, four members of Greenpeace snuck into Heathrow, strung a banner protesting a third runway onto the tail of a parked aircraft, and were subsequently arrested. At about the same time, 5 members of an anti-aviation group called Plane Stupid scaled the roof of the British Parliament and unfurled similar banners. The activists opposed plans to build not only a new terminal but also a new runway, which could allow Heathrow to expand from 480,000 flights a year to about 800,000.

During late November 2002, two official British studies called for an end to cheap flights and a ban on new airport runways. The Royal Commission on Environmental Pollution urged Britain's government to halt airport growth, raise fares, and place financial pressure on short-haul and no-frills carriers. At the same time, Prime Minister Blair's body of environmental advisers, the Sustainable Development Commission, said that proposals for new runways at Stansted, Heathrow, Luton, Rugby, or a new airport at Cliffe in Kent required a "fundamental rethink" (Clover and Millward 2002, 1). The commission, a body of academics, business-men, and people from public life, estimated that the price of a one-way ticket would need to rise by £40 to have any chance of mitigating climate change. Paul Ekins, an economist and a member of the commission, said, "We believe a stable climate is a good thing and worth modifying human behavior for" (Clover and Millward 2002, 1).

Some companies have been replacing air travel with teleconferencing. By 2007, GMAC Mortgage, a subsidiary of GMAC Financial Services, was using teleconferencing and Web conferencing for 75 percent of its train-ing, which was previously done by flying employees to a centralized loca-tion (Everson and Athavaley 2007, D-1, D-3).

As a postscript to the carbon footprint of aviation (and travel gener-ally), a "revolutionary idea" that no one will ever see advertised because no one makes any money from it, floated into my e-mail in-box, having been passed hand-to-electronic-hand several times: "The Simplest of Rev-olutionary Messages: STAY HOME!" "Avoid traveling vacations," the anonymous message said. "Keep the car in the garage except for essential tasks. Turn off the TV. Read your email but go take a walk once in a while! Get personally serious about that 'carbon footprint' idea" (Arden 2007).

The correspondent was animated:

> Hey, don't wait for this do-nothing government to alert you. That would slow down the profits. They and their corporate allies are the CAUSE of the disaster rapidly occurring around us known as "Global Warming." Do you need THEM to commit yourself to saving this Mother Earth? Hell NO!!! Create a local CANCEL THAT TRIP—STAY HOME THIS SUMMER group with your neighbors. Do you know them? It's time you did! Set up STAY HOME!! signs in the yard & ON telephone poles & talk to the folks next door. The survival of our species depends on YOU and THOSE AROUND YOU getting together. STAY HOME!! (Arden 2007)

## THE CARBON FOOTPRINT OF SCIENTIFIC MEETINGS

Scientific organizations have begun to take a hard look at the carbon emissions involved in their annual meetings, more than 90 percent of which involve air travel. The annual meeting of the American Geophysical Union (AGU), for example, held each fall in San Francisco, brought to-gether 9,500 people in 2002, who "spent" almost 11,000 metric tons of carbon dioxide traveling (Lester 2007, 36). Meetings of international cli-mate-change bodies have been severely criticized by deniers of global warming for flocking together on jets by the thousands as well.

The largest scientific meeting on Earth is the worldwide Society for Neuroscience, with about 35,000 people attending. The American Associ-ation for the Advancement of Science gathers about 8,000 a year (as of

2007). The Ecological Society of America has reduced its program and attendance out of concern over greenhouse emissions, and the AGU has been working on ways to webcast some of its sessions (Lester 2007, 36–37). Virtual meetings have been growing and may increase as better technology becomes more widely available. Scientists often do not want to give up the brainstorming and networking of face-to-face meetings, so one alternative is to look more closely at who attends and where they live and to schedule meetings at more centralized locations. Instead of jetting off to Australia, for example, a group whose members are mainly from the United States might meet in Chicago, Omaha, or Kansas City.

Some groups have offered carbon offsets, averaging about $20 for travel to the annual meeting, but only a tiny minority of attendees even at the Ecological Society of America have signed up for them: 500 of 3,600 registrants in 2007, or 15 percent (Lester 2007, 37).

## SHIPPING: SAILS REBORN

Some cargo ships that have been propelled solely by fossil fuels are using sails to reduce fuel consumption by as much as a third. One example is the M/V *Beluga SkySails* (Kleiner 2007, 272). The sails, produced in Bremen, Germany, by Beluga Shipping, cover as much as 5,000 square feet and can be manipulated to take advantage of wind direction, acting like parafoils that generate lift as well as propulsion—as much as a 5,000-kilowatt engine.

International shipping (like airlines) has long been exempt from greenhouse-gas limits, but some owners expect them to be imposed. They also are reacting to rapidly rising prices of oil and other fossil fuels. Cargo ships carry four hundred times as much freight by weight as airlines, four times as much as trucks, and six times the railroads' loads. The average ship carries a ton of cargo using one-fourth to half as much energy as these other modes of transportation, even without fuel-efficiency restrictions (Kleiner 2007, 272).

Shipping also can become more energy-efficient by reducing ships' speeds. Doubling the speed increases fuel consumption eight times. Attention also is being paid to more efficient hull and propeller designs, as well as a redesign of ships' engines. Improved hull designs alone may improve efficiency by as much as 15 percent (Kleiner 2007, 272).

California attorney-general Edmund G. Brown Jr. on October 5, 2007 joined three environmental groups (Oceana, Friends of the Earth, and the Center for Biological Diversity) in a request that the Environmental Protection Agency (EPA) issue and enforce greenhouse-gas regulations for cargo, cruise, and other commercial ships that use the oceans. The petition said that oceangoing vessels together emit more carbon dioxide than any single nation except the United States, Russia, China, Japan, India, and Germany. "Ominously, these emissions are projected to increase nearly 75 percent during the next 20 years," said Brown. "International law guarantees a right of 'innocent passage' for all ocean-going vessels, but this right does not include polluting the air or water near our coastal cities," Brown declared. "If the U.S. is to do its part in reducing the threat of global climate disruption, then EPA must limit the global warming emissions from ships that enter the ports of the United States" (Environment News Service 2007l).


## REFERENCES

Arden, Harvey. 2007. "STAY HOME!!" Email from harvey@harveysplace.net, July 24.

Clark, Andrew. 2006. "'Open Skies' Air Treaty Threat." *Guardian* (London), February 20. http://www.guardian.co.uk/frontpage/story/0,,1713677,00.html.

Clover, Charles. 1999. "Air Travel Is a Threat to Climate." *Daily Telegraph* (London), June 5.

Clover, Charles, and David Millward. 2002. "Future of Cheap Flights in Doubt; Ban New Runways and Raise Fares, Say Pollution Experts." *Daily Telegraph* (London), November 30, 1,4.

Daviss, Bennett. 2007. "Green Sky Thinking: Could Maverick Technologies Turn Aviation into an Eco-success Story? Yes, but Time Is Running Out." *New Scientist*, February 24, 32–38.

Environment News Service. 2006a. "British Travel Agents Launch Carbon Offset Scheme." November 28. http://www.ens-newswire.com/ens/nov2006/2006-11-28-05.asp.

———. 2007l. "EPA Petitioned to Limit Greenhouse Gases from Ships." October 5. http://www.ens-newswire.com/ens/oct2007/2007-10-05-094.asp.

———. 2007p. "Fueling Jets with Animal Fat." July 18. http://www.ens-newswire.com/ens/jul2007/2007-07-18-09.asp#anchor7.

Everson, Darren, and Anjali Athavaley. 2007. "The Downgrading of Business Travel." *Wall Street Journal*, July 3, D-1, D-3.

Fallows, James. 1999. "Turn Left at Cloud 109." *New York Times Sunday Magazine*, November 21, 84–89.

Flannery, Tim. 2005. *The Weather Markers: How Man Is Changing the Climate and What It Means for Life on Earth.* New York: Atlantic Monthly Press.

Francis, Justin. 2006. "Should the Responsible Traveller Be Flying?" Responsibletravel.com press release, February 10. http://www.responsibletravel.com/copy/copy900993.htm.

Frank, Robert. 2007. "Living Large While Being Green." Wealth Report. *Wall Street Journal*, August 24, W-2.

Friedl, Randall R. 1999. "Atmospheric Chemistry: Unraveling Aircraft Impacts." *Science* 286 (October 1): 57–58.

Hickman, Martin. 2006. "The Prince of Emissions." *Independent* (London), April 1, 1.

Higgins, Michelle. 2006. "Machu Picchu, without Roughing It." *New York Times*, August 12, Travel, 6.

Hillman, Mayer, Tina Fawcett, and Sudhir Chella Rajan. 2007. *The Suicidal Planet: How to Prevent Global Climate Catastrophe.* New York: St. Martin's/Dunne.

Houlder, Vanessa. 2002. "Rise Predicted in Aviation Carbon Dioxide Emissions." *Financial Times* (London), December 16, 2.

International Energy Agency. 1993. *Cars and Climate Change.* Paris: International Energy Agency.

Johnston, David Cay. 2000. "Some Need Hours to Start Another Day at the Office." *New York Times*, reprinted in *Omaha World-Herald*, February 6, 1-G.

Kleiner, Kurt. 2007. "The Shipping Forecast." *Nature* 449 (September 20): 272–73.

Kondratyev, Kirill, Vladimir F. Krapivin, and Costas A. Varotsos. 2004. *Global Carbon Cycle and Climate Change*. Berlin: Springer/Praxis.

Lean, Geoffrey. 2001. "We Regret to Inform You That the Flight to Malaga Is Destroying the Planet: Air Travel Is Fast Becoming One of the Biggest Causes of Global Warming." *Independent* (London), August 26, 23.

Lester, Benjamin. 2007. "Greening the Meeting." *Science* 318 (October 5): 36–38.

Lynas, Mark. 2006. "Fly and Be Damned." *New Statesman* (London), April 3, 12–15. http://www.newstatesman.com/200604030006.

McCrea, Steve. 1996. "Air Travel: Eco-tourism's Hidden Pollution." *San Diego Earth Times*, August. http://www.sdearthtimes.com/et0896/et0896s13.html.

McGuire, Bill. 2005. *Surviving Armageddon: Solutions for a Threatened Planet*. New York: Oxford University Press.

Michael, Daniel, and Susan Carey. 2006. "Airlines Feel Pressure as Pollution Fight Takes Off." *Wall Street Journal*, December 12, A-6.

Monbiot, George. 2006b. "We Are All Killers: Until We Stop Flying." *Guardian*, February 28. http://www.monbiot.com/archives/2006/02/28/we-are-all-killers.

Naik, Gautam. 2007a. "Arctic Becomes Tourism Hot Spot, but Is It Cool?" *Wall Street Journal*, September 24, A-1, A-12.

Reuters. 1999. "Aircraft Pollution Linked to Global Warming: Himalayan Glaciers Are Melting, with Possibly Disastrous Consequences." *Baltimore Sun*, June 13, 13-A.

Stoll, John D. 2006. "Visions of the Future: What Will the Car of Tomorrow Look Like? Perhaps Nothing Like the Car of Today." *Wall Street Journal*, April 17, R-8.

Stuber, Nicola, Piers Forster, Gaby Radel, and Keith Shine. 2006. "The Importance of the Diurnal and Annual Cycle of Air Traffic for Contrail Radiative Forcing." *Nature* 441 (June 15): 864–67.

*Wall Street Journal*. 2006. "Global Air Traffic Rose, Easing Fuel Cost Blow." July 2, A-7.

———. 2007. "Carbon Neutral Chic." Editorial. July 9, A-14.

Wilber, Del Quentin. 2007. "U.S. Airlines under Pressure to Fly Greener; Carriers Already Trying to Save Fuel as Europe Proposes Plan." *Washington Post*, July 28, D-1. http://www.washingtonpost.com/wp-dyn/content/article/2007/07/27/AR2007072702256_pf.html.

# CHAPTER 3

# Greening Shelter and Food

As 2007 dawned, amidst all sorts of apocalyptic warnings about a globally overwarmed future, corporate executives began "getting with the program." They were realizing, finally, that paradigm changes in energy use may be good for business. Suddenly, addressing global warming was not just a depressing litany of utter despair, of going hungry under a cold shower. New York City mayor Michael Bloomberg (himself enriched as an investor before he became a politician), sees a sea change, as "major business and financial institutions increasingly understand that shrinking the world's carbon footprint is a pro-growth strategy, indeed the *only* pro-growth strategy for the long term" (Revkin and Healy 2007).

As Henry Ford learned after his cars displaced thousands of blacksmiths, or as inventors of the first computers discovered from complaining accountants bent over hand ledgers, change makes money for those who find themselves (or put themselves) on the right side of history. Change often begins at home, in one's home or office. In this case, Bloomberg was among a number of political and business leaders newly united to retrofit old buildings in big cities for a new future of carbon-footprint awareness. On May 16, 2007, this group pledged billions of dollars' worth of investments to curtail urban energy use and thereby emissions of greenhouse gases under the initiative of the William J. Clinton Foundation.

Each participating bank committed to allocate as much as $1 billion for loans that city governments and private landlords may use to upgrade inefficient heating, cooling, and lighting systems in older buildings. The loans and interest would be repaid with savings realized through reduced energy costs. Bloomberg said that retrofitting older buildings was vital because 85 percent of them will still be standing (and occupied by energy-consuming human beings) several decades from now. The first targets under this

**Figure 3.1. Michael Bloomberg, mayor of New York City, delivering a speech in 2004.**
Source: Wikipedia.com. http://en.wikipedia.org/wiki/Image:Michael_Bloomberg_speech.jpeg.

initiative were the municipal buildings of the participating cities, including Bangkok, Berlin, Chicago, Houston, Johannesburg, Karachi, London, Melbourne, Mexico City, Mumbai, New York, Rome, São Paulo, Seoul, Tokyo, and Toronto (Revkin and Healy 2007).

## STREET-LEVEL SOLUTIONS

A third of human-generated greenhouse gases are produced in shelters—homes, manufacturing plants, offices. Reducing emissions at street level often can be embarrassingly easy. Once city governments and residents began to give the greening of urban America some serious thought, basic solutions began to pop out like flowers after a spring rain. During the early spring of 2007, I visited Scandinavia on assignment for the *Progressive* to describe what Sweden and Denmark were doing to move away from dependence on oil imports. I sat in a workroom at the Swedish Riksdag (Parliament) with Per Bolund, a Green Party member, who spent much of our time asking me how people in the United States were changing their carbon footprint. "You know," he said, "once the United States gets moving, it is marvelous to watch."

Escalators and moving sidewalks in the Riksdag stop when no one uses them. Many renovated Stockholm hotel rooms and apartments include a slot near the door for key cards that must be activated to turn on lights. When leaving a room, removal of the card turns off all the lights, making it impossible to leave a lit room. In 2007, Sweden began plans to label products and services for climatic impact.

A report by consultant McKinsey & Co. maintained that some of the most basic steps to curb global-warming emissions are the most cost-effective: improving energy efficiency of buildings, including lighting and air conditioning, and planting more trees. A report in the *Wall Street Journal* said, "Such moves will be cheaper than high-technology methods, such as capturing the $CO_2$ that is emitted from power plants and then burying that $CO_2$ underground" (Ball 2007b, A-12).

Even a small increase in a city's coverage of parks and trees along streets (as well as foliage-covered "green" roofs) may counter several degrees of anticipated temperature rise, according to a study at Britain's University of Manchester. The same parks and trees also would help retain rainwater from intensifying storms that are expected in a warming atmosphere that otherwise would drain away into streams and rivers, eventually returning to the sea (Environment News Service 2007e).

A 10 percent increase in the amount of green space in built-up city centers may reduce urban surface temperatures by as much as 4°C (7.2°F), according to this study. Temperatures would fall because of transpiration, the cooling of evaporating water from leaves and other vegetation.

"Green space collects and retains water much better than the built environment," explained Roland Ennos, a biomechanics expert in Manchester's Faculty of Life Sciences and a lead researcher in the team. "As this water evaporates from the leaves of plants and trees, it cools the surrounding air in a similar way to the cooling effect of perspiration as it evaporates from our skin" (Environment News Service 2007e).

A number of major cities have launched sizable tree-planting programs, including Washington, Baltimore, Minneapolis, Chicago, Denver, and Los Angeles. Still, the decline in tree cover has been accelerating since the 1970s, especially on private property and new development, according to American Forests, an environmental group that uses satellite imagery to document tree cover across the United States (Harden 2006, A-1). "This is like a creeping cancer," said Deborah Gangloff, the group's executive director. "In the two dozen cities we have studied, we have noticed about a 25 percent decline in tree canopy cover over the past 30 years. This is a dramatic trend that is costing cities billions of dollars" (Harden 2006, A-1). Washington, D.C., is among the cities with the largest reduction in dense tree cover, with a 64 percent decline from 1973 to 1997, according to American Forests.

Three shade trees strategically planted around a house can reduce home air-conditioning bills by about 30 percent in a hot, dry city such as Sacramento, California. About 375,000 shade trees have been given away to Sacramento city residents, and the city plans to plant at least four million more. To receive up to ten free trees, residents simply call the Sacramento Municipal Utility District (SMUD), a publicly owned utility company. "A week later, they are here to tell you where the trees should be planted and how to take care of them," said Arlene Willard, a retired welfare case worker who, with her husband John, has planted four SMUD trees in the backyard of their East Sacramento house (Harden 2006, A-1).

A nationwide shade program similar to the one in Sacramento could reduce air-conditioning use by at least 10 percent, according to Department of Energy (DOE) research (Harden 2006, A-1). A study of greenhouse-gas reduction potential in Chicago endorsed urban tree-planting projects to reduce air pollution in cities, where relief from urban heat is needed the most.

By planting 10 million trees as well as installing lighter-colored roofs and pavement, Los Angeles could reverse an urban "heat island" effect caused by concrete, asphalt, and heat-retaining buildings that has been increasing for a hundred years, according to a simulation study by the DOE's Lawrence Berkeley National Laboratory. It found that Los Angeles could lower its peak summertime temperature by five degrees, cut air-conditioning costs by 18 percent, and reduce smog by 12 percent (Harden 2006, A-1).

In 2006, Los Angeles started a campaign to plant a million trees, part of a free-tree program following the Sacramento model. For every dollar it spends on trees, the city expects to realize a $2.80 return from energy savings, pollution reduction, storm-water management, and increased property values, said Paula A. Daniels, a commissioner on the Board of Public Works (Harden 2006, A-1). Iowa in 2007 was the only state with a long-term record of using state law as an incentive for private utilities to plant trees for energy conservation.

## SIMPLE, BASIC ENERGY SAVERS

Other energy-conservation strategies related to shelter are as easy and obvious. Cut the amount of hot water consumed for showering in half instantly, for example, by using a method developed by the U.S. Navy on aircraft carriers with limited water supplies. Take a brief shower, then kill the water, soap up, and rinse. Sailors everywhere know the "Navy shower," and energy conservationists say it's time to adopt this simple solution on land.

One of the most basic energy-conservation strategies at home is an off-white roof that takes advantage of albedo, the scientific term for reflectivity. A light color reflects 80 percent or more of the sun's light and heat. A dark color reflects only about 20 percent. The difference in a home's summer cooling bill can be very noticeable.

Some of the people with the smallest carbon footprints in the United States live in tall buildings in the largest cities. New York City, for example, houses a large proportion of people who live without automobiles, taking subways and buses, walking, or riding bicycles. Their living spaces share walls with others—a key conserver of heat and air conditioning. Ironically, perhaps, a commune in the Vermont countryside usually carries a higher carbon cost than an apartment in New York City.

Installing motion sensors that dim the lights by 50 percent when hallways and stairwells are not in use could reduce the carbon dioxide emissions of a 60,000-square-foot building by about 40 metric tons, the equivalent of driving a 25-mile-per-gallon car 110,000 miles. Hiring an electrician to install the motion sensors would cost $10,000 to $12,000, according to estimates produced by Optimal Energy, a consulting company in Bristol, Vermont. The building could save that much money in lower electricity bills over two years, assuming that it uses fluorescent bulbs.

Other shelter-related energy savers are just as prosaic. Assigning a specific person in an office to switch off all lights and equipment at the end of an office day can cut carbon emissions by reducing electricity use, while also extending equipment life and reducing maintenance costs. Install motion-sensitive office lighting that extinguishes itself when no one is using a given space. Air conditioners and overhead lights can be timed to turn off; aim for off-peak energy use to be about one-fifth of peak use. In the morning, the switch-on monitor takes over (Masters 2007c).

At home, in the yard, gas-powered leaf blowers are the second-worst idea ever invented. The worst is the so-called patio heater, which delivers heat to open air. The United States has 30 million acres of front and back yards from which leaves can be removed just as easily with an old-fashioned rake (Masters 2007e). As is so often the case when one replaces fossil fuel with muscle power, exercise is an additional benefit.

U.S. consumers spend more than $5 billion each year on fertilizers manufactured from fossil fuels that leak toxic chemicals into the ground and accelerate the release of nitrous oxide, a powerful greenhouse gas. Compost and grass clippings can do as well or better. "More adventurous gardeners use a homemade fertilizer mix that includes seaweed extracts for potassium and fish proteins and oils for nitrogen. Or go native and embrace wildflowers and indigenous grasses. Weeds are a matter of taste" (Masters 2007d).

At the office, most white-collar workers can save a great deal of energy by thinking twice (or three times) before hitting the Print button. While we once were told that computers would make paper obsolete, the ease of printing has actually boosted paper consumption. People in the United States recycled 42 million tons of paper in 2006 (half of all the paper produced in the country), 900 million trees' worth. Recycled paper uses 60 percent less energy to manufacture than virgin paper (Masters 2007a).

Likewise, we can develop a taste for taking our products with a minimum of packaging. Traveling in Poland, I couldn't help but notice the small size of their garbage cans—or, comparatively, how large ours have become. Ours have been super-sized to accommodate a glut of paper and packaging that is sold to us along with the products we consume. A great deal of it is not necessary. All of this paper and plastic produces greenhouse gases in its manufacture. Every time you are offered packaging perform an on-site energy audit. Do you really need it?

Hewlett-Packard in 2007 switched to lighter packaging for its printer cartridges, which will reduce carbon emissions by an amount equivalent to removing 3,500 cars from the road for a year. Wal-Mart has taken the initiative to trim everything from its rotisserie-chicken boxes to its water bottles, cutting packaging 5 percent a year beginning in 2008, enough to prevent 667,000 tons of carbon dioxide emissions (Walsh 2007b).

At home (or the office), leaving household appliances on standby would not seem to cost much energy per unit, but adding up the bill for the 300 million people in the United States can be a shock. Alan Meier, a senior energy analyst at the Lawrence Berkeley National Laboratory, did just that, and estimated that U.S. households use 45 billion kilowatt-hours in standby electricity per year, at a cost of $3.5 billion, equal to the production of seventeen 500-megawatt power plants (Steinman 2007, 303–4). A study at the University of California at Berkeley and Lawrence Berkeley National Laboratory indicated that eliminating standby electricity loss from home appliances could produce substantial savings on electricity bills. The study found that standby usage ranged from 6 to 26 percent of homes' annual electricity use. An average of nineteen appliances per household were in standby mode (Sanders 2001).

The average desktop computer, not including its monitor, consumes between 60 and 250 watts per day. Turning a computer off after using it 4 hours a day could save about $70 in electricity costs per year. The carbon impact would be even greater. Shutting it off would reduce the machine's $CO_2$ emissions 83 percent, to just 140 pounds a year (Masters 2007f).

The household equivalent of a well-tuned car is a furnace that has been checked, cleaned, and, if it is more than a decade old, replaced. If everyone in the United States upgraded furnaces with the latest energy-efficient technology, natural gas use for space heating would probably fall at least 25 percent, according to the DOE. A new furnace may be installed as part of renovations to a basement.

In many parts of the United States (and in Great Britain as well), many urban dwellers can request that a portion of their electricity consumption come from renewable sources. The Green Power Network's website (http://eere.energy.gov/greenpower) lists utilities that offer power from alternative sources, usually for a fee. About 350 U.S. utilities offered such programs in 2007.

## CLOTHESLINE COMPLEXITIES

In the course of searching for simple, nearly immediate ways to shrink one's carbon footprint, I found one that sounded *really* easy: hang a clothesline, and use it. This is a simple, old-fashioned concept, but given the complexities of twenty-first-century life, the pursuit of sun-dried clothing poses unanticipated problems—enough, it turned out, to fill a very entertaining, lengthy article in the *New York Times* describing Kathleen Hughes's conversion from mechanical drying (Hughes 2007).

This tale began with Hughes's childhood, in Tamaqua, Pennsylvania, a small coal-mining town, helping her mother. It ends in Rolling Hills, California, a gated community that is described as being an hour's drive south of Los Angeles (depending on Los Angeles–area traffic conditions, that could be anywhere between 100 yards and 60 miles). Among undulating hills, ranch houses, sweeping views of the ocean, and rocky cliffs, Hughes could not find a single visible clothesline. In this atmosphere, Hughes's new clothesline became a novelty, but one with some appeal in a house with monthly electric bills as high as $1,120 in midsummer.

Soon Hughes discovered that the hanging of laundry in our time, once such a simple act, can be grounds for considering housing covenants and consulting "laundry activists" on the Worldwide Web. Don't we love the twenty-first century? Hughes found herself in "the laundry underground," a mixed group including "the frugal, people without dryers, and people from countries where hanging laundry is part of the culture" (Hughes 2007). Joining the clothesline underground in an affluent California gated community took some doing. Many people there associated clotheslines with visual pollution and poverty, not green consciousness. Hughes found the same to be true across the United States:

> Clotheslines are banned or restricted by many of the roughly 300,000 homeowners' associations that set rules for some 60 million people. When I called to ask, our Rolling Hills Community Association told me that my laundry had to be completely hidden in an enclosure approved by its board of directors. ... I'm supposed to submit a site plan of our property and a photograph of my laundry enclosure. But I don't have an enclosure, unless the hedge qualifies. Looking for fellow clothesline fans, I came across the Web site of Alexander Lee, a lawyer and 32-year-old clothesline activist in Concord, N.H. In 1995 Mr. Lee founded Project Laundry List, a non-profit organization, as a way to champion "the right to dry." His Web site, laundrylist.org, is an encyclopedia on the energy advantages of hanging laundry. ... Mr. Lee sponsors an annual National Hanging Out Day on April 19. (Hughes 2007)

Think what Hughes would have had to do to install a wind turbine in her backyard! She also encountered problems at home, where her husband complained that while sun-dried towels smelled fresh, they were abrasive: "'Like sandpaper,' said my husband, Dan, after stepping out of the shower" (Hughes 2007). Give Dan a Purple Heart in the Battle to Squelch Greenhouse Gases.

Very close to home, in Omaha (and not in a gated community), my wife and a neighbor to the west have been hanging clothes to dry. Sizing up her clothesline with a look of disgust, a neighbor across the fence to the south asked our neighbor whether her dryer was broken. When our neighbor said she was drying with solar energy to save electricity and impede global warming, she was told, pointedly, that hanging clothes in the sun was an offensive form of visual pollution.

Hughes calculated: "There were more than 88 million dryers in the country in 2005, the latest count, according to the Association of Home Appliance Manufacturers. If all Americans line-dried for just half a year, it would save 3.3 percent of the country's total residential output of carbon dioxide" (2007). In the meantime, back in her mother's hometown, the old clotheslines never came down.

## BUILDING CODE CHANGES AROUND THE WORLD

Considerable mitigation of global warming may be accomplished by wise use of new technology to improve the energy efficiency of dwellings, factories, and offices. Energy consumption of heating and air-conditioning systems could be reduced by as much as 90 percent in new buildings, for example, with modern insulation, triple-glazed windows with tight seals, and passive solar design (Speth 2004, 65). Ceramic powder can even be mixed with interior house paint to reduce the amount of hot or cold air seeping through walls.

On January 1, 2003, Australia changed its national building code with the explicit purpose of reducing energy consumption. Amendment 12 of Australia's building code requires wall, ceiling, and floor insulation to avoid or reduce the use of energy for artificial heating and cooling, utilizing passive solar heating where appropriate, as well as natural ventilation and internal air movement. The code requires sealing houses in some climates to reduce energy loss through leakage and insulation to reduce heat loss from water piping of central heating systems, along with insulation and sealing to reduce energy loss through the walls of ductwork associated with heating and air-conditioning systems.

British Columbia initiated a "Green Building Code" during 2007 containing incentives to retrofit existing homes and buildings for energy efficiency. The code requires energy audits, as well as plans for "real-time, in-home smart metering" to help homeowners measure and reduce energy consumption. Separate strategies promote energy conservation in universities, colleges, hospitals, schools, prisons, ferries, and airports (Environment News Service 2007d).

On a brownfield (formerly toxic) site in London's Docklands, construction is under way for a 2010 opening of the city's first large-scale zero-carbon housing development, with 233 homes that will share a combined heat-and-power plant that uses wood chips to create electricity and heat water, with extra input from solar panels and wind. The residents will use the regular power grid only if their own homegrown sources do not meet the demand. The London "Greenzone" is also designed to accommodate greenhouse-grown organic food and car and bicycle clubs to reduce greenhouse emissions.

Danish building codes enacted in 1979 (and tightened several times since) also require thick home insulation and tightly sealed windows. Between 1975 and 2001, Denmark's national heating bill fell 20 percent, even as the amount of heated space increased by 30 percent. Denmark's gross domestic product has doubled on stable energy usage during the last thirty years. The average Dane now uses 6,000 kilowatt-hours (kwh) of electricity a year, less than half of the U.S. average (13,300 kwh). Simply utilizing waste heat ("cogeneration") can save a substantial amount of that wasted power, with existing, tested technology.

Denmark by 2006 was obtaining half of its power through distributed grids, as carbon dioxide emissions fell from 937 grams per kilowatt-hour in 1990 to 517 in 2005 (Butler 2007, 587). Surplus heat from Danish power plants is piped to nearby homes, via insulated pipes, for example, using cogeneration or "district heating," which required tearing up streets to install the pipes. Power plants were downsized and built closer to people's homes and offices. In the mid-1970s, Denmark had fifteen large power plants; it now has several hundred. By 2007, six in ten Danish homes were heated this way, which is less expensive than oil or gas (Abboud 2007, A-13).

## SUSTAINABILITY STANDARDS FOR NEW CONSTRUCTION

The United States has no federal regulations that require sustainability in new construction. Private Leadership in Energy and Environmental Design (LEED) guidelines, drawn up by the U.S. Green Building Council, a nonprofit group founded in 1993, suggest voluntary compliance. These have been adopted by several government agencies, including the General Services Administration, which oversees the construction of federal buildings. In Cincinnati, Ohio, residents who build or remodel according to LEED standards are excused from paying city property taxes for fifteen years. Even without regulation, some architectural engineers are utilizing sustainable principles in their designs. The new Hearst Tower in Manhattan has natural ventilation and high-performance glass that deflects heat (Ouroussoff 2007).

Until 2007, LEED guidelines applied only to public and private buildings. In November of that year, the Green Building Council issued similar guidelines for private residences. A "green" house may contain solar panels, several types of insulation, a roof garden that filters rainwater for irrigation, thermal windows and doors, a geothermal heat pump, and other design features.

Buildings certified in accordance with LEED guidelines produce an average of 38 percent less carbon dioxide and reduce energy use 30 to 50 percent compared to U.S. averages. The organization's intricate certification process can cost as much as $30,000, but the cost is usually recovered in a few years of energy savings (Jordan 2007a, D-2). LEED rates buildings on a four-tier scale, based on energy savings and payback period: *certified* (energy savings of 25 to 35 percent, with a payback period less than three years), *silver* (35 to 50 percent and three to five years), *gold* (50 to 60 percent and five to ten years), and at the top, *platinum* (more than 60 percent and more than ten years). By 2007, 10,000 organizations had memberships in LEED, along with 91,000 individuals—contractors, developers, architects, building owners, universities, suppliers, and others (Jordan 2007a, D-1; U.S. Green Building Council 2008).

The Green Building Council even certifies mega-mansions. A 7,000-square-foot home in Atlanta owned by Laura Turner Seydel, daughter of Ted Turner, and her husband even has a name ("EcoManor"), earned with its

27 photovoltaic panels on the roof, rainwater-collecting tanks for supplying toilet water, and "gray water" systems that use water from the showers and sinks

for the lawn and gardens. ... The home has ... a switch near the door that turns off every light in the house before the family leaves. (Frank 2007, W-2)

Seydel asserts that her energy bill is half that of the usual 7,000-square-foot house.

Architects have been finding all sorts of ways to reduce energy use: buildings are being constructed with an abundance of windows that allow passive solar energy to reduce electricity use; overhangs on exterior walls screen the sun in the summer and allow it in in winter; walls of limestone on the sunny sides of buildings block the heat and reduce air-conditioning use. The annual Greenbuild conference for developers, architects, contractors, suppliers, and others drew 4,200 people to Austin, Texas, in 2002; that number rose by a factor of four to about 18,000 in Chicago in 2007. By that year, "green" construction was a $12-billion-a-year market (Jordan 2007a, D-2).

Some cities are now requiring green design in new and renovated large commercial buildings. A San Francisco Green Building Task Force report issued in July 2007 recommends requiring an immediate target of LEED certified, increasing to a more stringent standard of LEED gold by 2012. For smaller commercial buildings, where the payback for green design is less substantial, the task force recommends voluntary compliance.

## ENERGY EFFICIENCY SAVINGS WITH EXISTING TECHNOLOGY

A study by the McKinsey Global Institute

> concludes that projected electricity consumption in residential buildings in the United States in 2020 could be reduced by more than a third if compact fluorescent light bulbs and an array of other high-efficiency options including water heaters, kitchen appliances, room-insulation materials and standby power were adopted across the nation. Energy conservation over that time, if achieved, would be equivalent to the production from 110 new coal-fired 600-megawatt power plants. (Lohr 2007)

The study recommends market intervention to correct distortions. A landlord, for example, has little incentive to upgrade inefficient appliances, because tenants usually pay the power bills. Tenants, who do not own properties, have no economic reason to purchase efficient appliances. "Everyone would be better off if the capital investments were made," said Diana Farrell, director of the institute, McKinsey & Company's economics research arm. "But this would be market intervention to correct market distortions that exist" (Lohr 2007). When individual parties do not have incentives to make the needed investments, this report suggests that more stringent product standards require that all new appliances be energy-efficient (Lohr 2007).

The McKinsey report's ideas are largely borrowed from California, which, beginning during the 1970s, established requirements for appliances and building materials, among other energy-saving measures. From

1976 to 2005, per-capita electricity consumption remained stable in California as it grew 60 percent in the rest of the United States (Romm 2007, 23). California, with its increasing population and growing high-tech economy, changed its utilities' pricing structure so that profits were not tied directly to how much power they sold. Instead, the state's electricity regulators decided to reward efficiency, in a program that has since lowered the energy bills of Californians by $12 billion a year ($1,000 per family) and avoided emission of more than 10 million metric tons of carbon dioxide per year (Romm 2007, 164). This provides a cogent example of what can be done in the United States as a whole, where one-third of the fossil fuels used to generate electricity annually is lost in generation or transport—more total energy than Japan uses each year.

## UNIVERSITIES GO CARBON NEUTRAL

By 2007, several dozen colleges and universities across the United States had pledged to achieve "climate neutrality" on their campuses. The American College and University Presidents' Climate Commitment is modeled on a similar agreement by city mayors across the United States (Environment News Service 2007k). Signers include chief executives of many small schools and some larger ones, such as the 200,000-student University of California (UC) system; the University of Florida, which has more than 50,000 students; the 29,000-student University of Colorado–Boulder; and the University of Hawaii at Manoa, with 20,600 students (Environment News Service 2007k).

These universities have joined the Association for the Advancement of Sustainability in Higher Education, ecoAmerica, and Second Nature in plans to neutralize their effects on climate within two years. The universities also agreed to help reduce their emissions of greenhouse gases by beginning at least two of the following six actions:

1. Establish a policy that all new campus construction will be built to at least the U.S. Green Building Council's LEED Silver standard or equivalent.
2. Adopt an energy-efficient appliance purchasing policy requiring purchase of U.S. federal government Energy Star-certified products in all areas for which such ratings exist.
3. Establish a policy of offsetting all greenhouse gas emissions generated by air travel paid for by our institution.
4. Encourage use of and provide access to public transportation for all faculty, staff, students and visitors at our institution.
5. Within one year of signing this document, begin purchasing or producing at least 15 percent of our institution's electricity consumption from renewable sources.
6. Establish a policy or a committee that supports climate and sustainability shareholder proposals at companies where our institution's endowment is invested. (Environment News Service 2007k).

The UC system in 2007 adopted its own sustainability policy that covers energy use and conservation, global warming, waste recycling, green building standards, and eco-friendly purchasing for its ten campuses. It thus became the first in the nation to adopt specific guidelines regarding purchase of electronics, disposal of waste, and recycling (Environment News Service 2007z). The university buys more than ten thousand computers a month and disposes of about a million pounds of waste each year (Environment News Service 2007z).

The university system's policy stemmed in part from a student-run "Toxic Free UC" campaign sponsored by the Silicon Valley Toxics Coalition. "With this new policy, UC and UC students can use their purchasing power to move electronics companies to make greener products that are less toxic and more easily recyclable. The UC is truly taking the lead toward a more sustainable future," said Maureen Cane, Silicon Valley Toxics Coalition's campus organizer. Under its new policy, UC pledged to buy only products registered under the Electronic Product Environmental Assessment Tool (EPEAT), developed by the Green Electronics Council (Environment News Service 2007z). EPEAT monitors computer equipment to environmental standards, requiring reduction of harmful chemicals and recycling, including standards that extend the life of computers that have been replaced every two to three years, making them a major source of waste.

## THE INTERNATIONAL BUILDERS' SHOW GOES GREEN

Each year, the International Builders' Show features a "new American home." In 2007, the group highlighted energy efficiency and green design. At 4,707 square feet, with a price tag of $5.3 million, this is not an average home. It is meant, rather, as a source of ideas for builders and designers of homes in the future. The house has a high-performance heating and cooling system to serve each of its three floors, with the system and ductwork entirely within the space that is being heated and cooled, to minimize waste. A solar thermal hot-water system feeds into a "tankless" system. Rooftop solar panels provide an average of an extra nine kilowatt-hours a day. Exterior walls are eight-inch precast insulated concrete sandwich panels that also minimize loss of warm or cool air. Ed Binkley, a partner in Bloodgood Sharp Buster Architects and Planners of Oviedo, Florida, who designed the home, said, "You'd have to take a crane with a wrecking ball to knock these walls down" (Rogers 2007, RE-2).

Windows on the south and west sides of this home are shaded with four-foot overhangs that include roof vegetation, part of a "green roof" that employs plants for insulation. The house is designed to use 73 percent less energy for heating and cooling and 54 percent less energy for heating water, compared to a conventional house with the same amount of floor space (Rogers 2007, RE-2). A cistern is used to collect all drainage from the roof and sides of the house, as well as runoff from air conditioners' condensers, for reuse in landscaping.

## HOTELS GO GREEN

King Pacific Lodge, a luxury resort in the Great Bear rain forest along the British Columbia coast (where a three-night stay costs about $5,000 per person), has turned down the temperature of its showers and fills its twin-engine boats (usually used for salmon fishing) to capacity and restricts their speed with engine governors, all to curtail greenhouse-gas emissions. The lodge is acting on a plan to cut its carbon footprint by half in five years (Ball 2007a, P-1). The lodge's carbon footprint had been enormous—1.7 tons per guest per stay (3.5 days), or about as much as an average U.S. citizen generates in a month. The resort's biggest carbon producers are two 110-kilowatt diesel generators for electricity. Propane is being substituted in the kitchen.

The most carbon-intensive part of a vacation in a remote area is usually the air travel required to reach it. The *Wall Street Journal* carried an account of one couple who installed a solar hot-water heater in their home and, feeling environmentally correct, took flights around the world. Soneva Fushi, a sixty-five-villa resort in the Maldives, a chain of islands in the Indian Ocean (often mentioned as one of the first locations that will drown in rising seas generated by melting ice), announced plans to go carbon neutral by 2010, partially by using coconut oil in its diesel generator. Devices also were installed to cut off the air conditioning if a room's door was left open for more than a minute (Ball 2007a, P-5). In addition, the resort added a fee to guests' bills to offset the massive amount of carbon dioxide generated by getting there by air from anywhere else.

By 2007, many business guests were demanding environmental consciousness from their hotels, and owners discovered that green decisions could save money—light bulbs that use less energy and bathroom fixtures that limit water flow saves money in the long run. "Environmental issues are one of the hottest issues within the travel industry right now," said Bill Connors, the executive director of the National Business Travel Association, which made eco-friendly elements in hotel design and operations a focus (for the first time) at its annual convention in July 2007 (White 2007).

Increasing numbers of hotels have been registering for certification under the U.S. Green Building Council's LEED program. Compact fluorescent bulbs, policies that extend the use of towels between washings, green roofs, use of nontoxic cleaning agents, and recycling bins in rooms have become more popular. Marriott Hotels plans to reduce energy consumption in its hotels 20 percent between 2000 and 2010.

Guests who drive hybrid vehicles to the Fairmont hotels in California and British Columbia receive free parking. The Habitat Suites in Austin installed a solar hot-water heating system that cut natural gas use 60 percent. The Lenox Hotel of Boston composts 120 tons a year of restaurant waste (Bly 2007, D-2). At San Francisco's Kimpton Hotel Triton, guests can book on a designated "Eco Floor," with low-flow toilets and nontoxic cleaning supplies. Shampoo and conditioner are dispensed from wall-mounted canisters, instead of individual plastic bottles.

At Greenhouse 26, to open during 2008 in Manhattan, the elevator will capture the energy generated when it stops, much as a hybrid car recycles energy released when the brakes operate. Heating and cooling will be provided geothermally. Water from sinks and showers will be recycled for use in toilets (White 2007).

Key card systems that control lights are now being used in many hotels. Room lights turn on when a guest inserts a card next to the door, and turn off when it is removed. It is therefore impossible to leave lights on in an empty room. Such systems are becoming common in several countries, including Sweden as mentioned earlier. Along similar lines, some hotels now have infrared sensors that turn off air conditioning or heat when a guest leaves a room. The cost of such a system usually can be recouped in a year or two from energy savings.

## THE ATMOSPHERIC CHEMISTRY OF AIR CONDITIONING

Air conditioning (especially the window-mounted variety) is a gigantic contributor to global warming based on its massive appetite for electrical energy. That's not all, however. The most popular refrigerant for air conditioners, hydrochlorofluorocarbons (HCFCs, specifically HCFC-22), also contributes to global warming. International pressure has grown rapidly for quick action to regulate HCFC-22. "We scientifically have proof: if we accelerate the phaseout of HCFC, we are going to make a great contribution to [mitigation of] climate change," said Romina Picolotti, the chief of Argentina's environmental secretariat (Bradsher 2007). Representatives of 191 countries that are parties to the Montreal Protocol met on the same site on the twentieth anniversary of that agreement and agreed unanimously September 21, 2007, to accelerate phaseouts of HCFCs, beginning during 2009.

On a molecular level, a derivative of HCFC-22 is about 11,700 times as potent as carbon dioxide at absorbing heat in the atmosphere. Phasing it out could cut emissions of global-warming gases by at least one-sixth of the amount required under the Kyoto Protocol (Bradsher 2007). The 1987 Montreal Protocol that restricted ozone-depleting gases contained an exception for HCFC-22 production by developing countries. China and India, two of these countries, have since become leading manufacturers of air-conditioning systems that use it. In addition to manufacturing, air-conditioning use has been doubling every three years in China and India as their economies boom. The same is true in other countries. Use of HCFC-22 increased a hundredfold in the African country of Mauritius between 2000 and 2005 due in large part to luxury hotel construction and rapid expansion of a fishing industry (with large exports) that uses refrigeration (Bradsher 2007).

The Kyoto Protocol exempted HCFC-22 and other ozone-depleting substances on the grounds that the Montreal agreement had already addressed the issue (Bradsher 2007). The 1987 protocol allows developing countries to increase production of HCFC-22 until 2016 and then freezes production at that level until 2040, with a phaseout following. That timetable was drawn up in the early 1990s, when HCFC-22 was used mainly in industrial nations; developing countries were then believed to be too poor to afford it. Industrialized countries (as defined in 1987) were required to phase out HCFC-22 manufacture by 2020. The European Union banned them completely in 2004. The United States plans a ban in 2010. Many European and American air conditioners are now imported from India and China, using more modern (and expensive) refrigerating agents.

In a twist on its usual lack of environmental priorities, the Bush administration by 2007 was strongly supporting a worldwide phaseout of HCFC-22, with lobbying muscle from DuPont, the giant chemical company, which stands to profit handsomely from manufacture and sale of a replacement, hydrofluorocarbons, which contain no chlorine and thus minimize impact on both ozone depletion and global warming. The United States, while resisting international diplomatic efforts to reduce greenhouse-gas emissions on the grounds that they will harm its economy, was beating the diplomatic drum to advance the phaseout date for HCFC-22 to 2020 from 2030 for industrial nations, and 2030 from 2040 for developing countries (Fialka 2007b, A-8).

## ELECTRIC METERS THAT RUN BOTH WAYS

The traditional electricity meter has entered the digital age at nearly the same time that energy infrastructure is changing to allow individual households to contribute their own power with sun, wind, and biogas, making electricity generation a two-way street. A centralized electricity-generation grid that relies on fossil fuels (meanwhile losing half of what it produces as waste heat and another 8 percent to transportation inefficiencies) will be replaced within decades by a system that will allow consumers to produce power and sell their surpluses back to the grid at market rates. Enel, an Italian electricity company, already had distributed 30 million "smart" meters between 2001 and 2006. By 2011, Pacific Gas and Electric, which serves northern and central California, plans to have supplied several million customers with smart meters (Butler 2007, 587).

As of 2007, forty of the fifty U.S. states had laws requiring "net metering," which provides the legal infrastructure for individual households to generate power and, if they have a surplus, sell it into the power grid at market rates, usually about 6 to 8 cents per kilowatt-hour. Wind power can generate electricity at that cost. Solar power (at 25 to 35 cents) is still too expensive to be competitive, but Concentrating Solar Power (CSP) may change that. By 2007, it was generating solar power at 9 to 12 cents per kilowatt-hour before subsidies.

Technological improvements in solar technology also may bring down the cost. Some farmers have been installing devices that turn hog manure into methane gas, a double winner because the process not only produces energy but also removes a greenhouse gas from the atmosphere.

There is every reason to expect that nonfossil fuels will become more economical in the future. Wind power, for example, cost an average of 80 cents per kilowatt-hour in 1980. Technological innovations and increasing demand cut that average to 10 cents in 1991, and 6 to 8 cents (both assuming good wind conditions) by 2007. At other sites, the cost of generation can range up to 29 cents per kilowatt-hour. Materials used in wind turbines have improved, and they are now larger and more efficient. Rotor diameter now ranges up to 400 feet (longer than a football field), compared to about 35 feet during the 1970s.

## GEOTHERMAL ENERGY: NEWLY POPULAR AT STREET LEVEL

In Iceland, 90 percent of the country's 290,000 people were using geothermal energy to heat their homes by 2007. Nearly all new buildings in Iceland are constructed with geothermal sources built in. By 2006, only 0.1 percent of Iceland's electricity came from fossil fuels, even as it attracted energy-intensive industries such as aluminum smelting drawn by low-cost electricity from renewable sources. Within one generation, Iceland transformed its energy system from a fossil-fuel base to one in which 70 percent of all energy (including transport) comes from renewable sources.

Large-scale geothermal plants are being constructed in the United States as well. By 2006, the United States was the world's largest producer of geothermal electricity—212 plants and 3,119 megawatts. The total is growing. During September 2007, Iceland America Energy started drilling just west of California's Salton Sea to build a geothermal power plant to supply Pacific Gas and Electric with 49 megawatts of electricity by 2010 (Eilperin 2007a, A-1). Even so, geothermal sources in 2006 accounted for barely half of 1 percent of U.S. energy generation.

Geothermal energy could produce 10 percent of U.S. electricity by 2050, according to a report by a team at the Massachusetts Institute of Technology. In 2007, geothermal was feasible at 6 to 10 cents per kilowatt-hour, without subsidies. Harrison Elementary School in Omaha remodeled with it; this is one example of several just in the parts of Omaha I can reach on my bicycle. Geothermal has suddenly become very popular at street level here in mid-America. As I watched the earth being turned over to save energy, I could not help but think of how old ideas have become new: Native Americans in my home state took advantage of geothermal energy with "earth lodges" long ago, and the earliest European-American "sodbusters" emulated them, burrowing into the prairies and plains, surviving, at least for a time, in housing that had no mechanical heating or cooling in a very rugged climate.

Warm springs are not required for geothermal energy. Thermal contrast between earth and atmosphere is enough, especially in places such as Omaha with large contrast in seasonal temperatures. The earth's temperature below the surface is about 56°F year-round; by circulating water through pipes above and below ground, as much as 70 percent can be saved on heating and cooling costs. When the air temperature is close to that of the earth, the need is minimal; the greater the contrast, the greater the need, and the more energy is conserved.

The system uses underground pipes filled with fluid that pull heat from buildings in the summer and release it into underground soil. In winter, the pipes distribute heat in the building that has been gathered underground. The principle is the same as that used by traditional residential heat pumps, but 30 to 50 percent more efficient (it uses 30 to 50 percent

**Figure 3.2. Geothermal heat pump equipment at Beijing Concordia International Apartment Building.**
Source: National Energy Laboratory, Photographic Information Exchange. U.S. Department of Energy.

less energy per unit of heat). The geothermal pumps use the ground, whereas the residential pumps use the air. Nationwide, installation of geothermal pumps has been growing at double-digit rates, according to John Kelly, executive director of the Geothermal Heat-Pump Consortium in Washington, D.C. (Gaarder 2007a, A-2).

Installation of a geothermal system involves major infrastructure costs. In addition to the one at Harrison Elementary, our local grade school, I watched a similar system being installed at St. Margaret Mary's Catholic School across Dodge Street from our campus. A soccer field was plowed up to install the pipes, part of a $2.3 million project that will recoup expenses in nine to thirteen years. (A residential heat pump, without the pipe work, cost $5,000 to $7,000 with a payback of eight to ten years). Such projects are feasible for schools and other institutions that will occupy the same property for decades.

## COMPACT FLUORESCENT LIGHT BULBS

Australia's government has required a nationwide phaseout of incandescent light bulbs in favor of compact fluorescent lights by 2010. While up to 90 percent of the energy from incandescent bulbs is wasted, mainly as heat, a compact fluorescent uses about 20 percent as much electricity to produce the same amount of light. A compact fluorescent light bulb can last between four and ten times longer than the average incandescent.

The new policy, announced by Environment Minister Malcolm Turnbull during 2007, should reduce Australia's greenhouse-gas emissions by four million tons within two years. Household lighting costs could be reduced by up to 66 percent, Turnbull said. "The most effective and immediate way we can reduce greenhouse-gas emissions is by using energy more efficiently," Turnbull said. (Environment News Service 2007a). In Australia, lighting currently represents around 12 percent of greenhouse-gas emissions from households and about 25 percent of emissions from commercial businesses (Environment News Service 2007a).

Australian Green Party members said that the government was taking mere baby steps toward the kind of energy conservation required to seriously challenge global warming. Screwing in new light bulbs will not do it, they said. The Greens aimed to stop coal mining, the source of 85 percent of Australia's electricity. "The proposed Anvil Hill coal mine in New South Wales will generate 28 million tons of carbon dioxide emissions every year, whereas the light bulb change will reduce emissions by 800,000 tons per annum," the Greens said. Wearing "Stop Anvil Hill" banners, the activists warned that the New South Wales government has plans to approve or expand at least eight mines in the Hunter Valley, which they said would be a "climate change disaster" (Environment News Service 2007a).

## ECO-CITIES IN CHINA AND DENMARK

In a country that by 2007 was industrializing mainly with power generated by burning dirty coal, with sixteen of the world's twenty worst air-polluted cities, China is planning an "eco-city" on the island of Chongming (meaning "Lofty Brilliance") at the mouth of the Yangtze River, an hour by ferry from Shanghai. The city is slated to open in 2010 with twenty-five thousand inhabitants. In Dongtan ("Eastern Bank"), the first of three bedroom communities that will be linked to Shanghai by a 15.6-mile bridge-tunnel, energy will be provided by solar and wind power and biofuels, including recycled organic material, down to household vegetable peels. Grasses will be planted on rooftops for insulation, and a quarter of the land will be reserved in its natural state as an ecological buffer. The village aims to be carbon neutral and to produce very little waste; it will have no landfill, and sewage will be processed for irrigation and composting (MacLeod 2007, 9-A).

China's experiment has been prompted by the environmental ravages of rapid industrial development that include pollution of all its major rivers, dangerously polluted urban air, acid rain over a third of its landmass, and desertification that has been claiming an area the size of Connecticut in an average year. The urban area of Dongtan will be designed for walking, public transit, and bicycles, with private cars discouraged. Plans call for the initial population of 25,000 to grow to 80,000 by 2020 and 500,000 by 2030—that is, if rising seas provoked by global warming do not swamp the island, which also may be an easy target for occasional typhoons. Construction began in September 2007. Yang Ailun, a climate and energy specialist with Greenpeace in Canada, said that the eco-city could "disappear because of climate change" within decades (MacLeod 2007, 9-A).

Meanwhile, in Denmark, the four thousand residents of largely agricultural Samso Island accepted a challenge from the Danish government during the 1990s to convert to a carbon-neutral style of life. By 2007, they had largely accomplished the task, with no notable sacrifice of comfort. Farmers grow rapeseed and use the oil to power their machinery; home-grown straw is used to power centralized home-heating plants; solar panels heat water, which is stored for use on cloudy days (of which the 40-square-mile island has many); and wind power provides electricity from turbines in which most families on the island have a share of ownership. As of 2007, the people of the island were building more wind turbines to export power.

## EATING LOW ON THE FOOD CHAIN

My wife, Pat Keiffer, quit eating beef after she taught for a time at Omaha Metro Community College's South Campus, which abuts a large slaughterhouse. Many days as she drove to campus, Pat's path was crossed by railcars full of doe-eyed young cattle with terrified looks on their faces that had Pat, who is partially Jewish, thinking of trains pulling into Auschwitz-Birkenau. The fact that beef is a greenhouse-gas-intensive food just added to her determination to avoid eating it in any form. Food production in the United States today requires six calories of energy to produce one calorie of food. Much of this is expended in transport, but also in energy-intensive cultivation of factory farms, as well as many people's preference for energy-intensive food such as meats, most notably beef (Hillman, Fawcett, and Rajan 2007, 60).

During late November 2006, the United Nations Food and Agriculture Organization stated that the livestock business generates more greenhouse-gas emissions than all forms of transportation combined. "Environmentalists are still pointing their fingers at Hummers and SUVs when they should be pointing at the dinner plate," said Matt A. Prescott, manager of vegan campaigns for People for the Ethical Treatment of Animals (PETA) (Deutsch 2007b). To make the point, PETA acquired a Hummer and draped over it a vinyl banner naming meat as the top cause of global warming. The Hummer, with a driver in a chicken suit behind the wheel, is now a mobile billboard for the carbon footprint of animal protein. "You just cannot be a meat-eating environmentalist," said Prescott.

The Humane Society of the United States ran advertising in environmental magazines on the same theme. The ads depict a car key and a fork. "Which one of these contributes more to global warming?" the ads ask. They answer the question with, "It's not the one that starts a car" (Deutsch 2007b). Judging by statistics, the folks at PETA have some work to do. The average American today eats more than 200 pounds of red meat, poultry, and fish per year, an increase of 23 pounds compared to 1970.

## THE CARBON FOOTPRINT OF MEAT

Many people are now going vegetarian as part of a personal climate-change strategy. Billions of chickens, cattle, turkeys, pigs, and cows being raised in factory farms produce methane from digestion and feces. The raising of animals for human consumption is the single largest source of methane emissions in the United States, and a methane molecule is twenty times as effective at retaining heat as one of carbon dioxide.

Animal agriculture is also a major source of carbon dioxide. Production of animal protein requires about ten times the fossil-fuel input (producing ten times the carbon dioxide) compared to edible plants. The feeding, killing, and processing of meat is enormously energy intensive. Taking such things into account, a study by University of Chicago professors Gidon Eshel and Pamela Martin argues that a completely vegetarian diet (avoiding eggs and dairy products as well as meat, fish, and fowl) can remove as much carbon dioxide and methane from the air (or, in some cases, more) than driving a hybrid car (compared to one with an internal combustion engine). They calculated that the difference between a meat-centered diet and vegetarianism had the same effect on greenhouse gas production as switching from an SUV to a standard sedan. Nitrous oxide in manure (warming effect: 296 times greater per molecule than that of carbon dioxide) and methane from animal flatulence (23 times greater) mean that "a 16-oz. T-bone is like a Hummer on a plate" (Will 2007, A-27).

Then there's Ben & Jerry's ice cream, invented by eco-talking Vermont hippies, a gallon of which

> requires electricity-guzzling refrigeration from manufacture to table, as well as four gallons of milk from cows that, at the same time, produce eight gallons of manure and eight gallons' worth of methane flatulence. The cows eat grain and hay cultivated with tractor fuel, chemical fertilizers, herbicides and insecticides, transported by trains and trucks. (Will 2007, A-27)

## FAR-FETCHED FOOD

A walk through an average U.S. supermarket is a tour of the world, a cornucopia of food choices. The carbon-conscious sometimes take pride in eating low on the food chain, substituting vegetables and fruits for meat. What, however, if our fruits and vegetables come from Chile, Nicaragua, and New Zealand out of season? In that case, most of their carbon calories come in the form of transportation, the majority of it powered by oil. The choices are beguiling. How long has it been since we could indulge our taste for watermelon only during late summer? I am just climatically careless enough to fall for Chilean grapes and watermelon from Nicaragua in January. I would rather not calculate the carbon footprint of that little "personal" Nicaraguan watermelon.

Food sold in U.S. grocery stores travels an average of 1,500 miles to reach consumers, according to David Pimentel, professor of ecology and agricultural science at Cornell University. The food industry burns almost one-fifth of all the petroleum consumed in the United States, about as much as automobiles, according to Michael Pollan's *Omnivore's Dilemma* (Knoblauch 2007, 46).

A tour of a typical U.S. supermarket also reveals that modern transport and various free-trade agreements—the North American Free Trade Agreement (NAFTA) and others—have rendered distance nearly irrelevant to our food supply. The idea of "seasonal" produce has nearly lost its meaning. During winter in the United States, many fruits and vegetables are now routinely imported from the Southern Hemisphere. Geography (and energy expenditure) has become so irrelevant in today's American grocery stores that some salmon are raised in the Pacific Northwest, shipped to China to be cut (where labor is cheap), and shipped back to North America to be sold.

Latino and Asian immigration to the United States has given rise to all manner of imports (packaged goods, mainly) from these countries— Mexico southward, China, Korea, Vietnam, India, and others. A typical city of a half-million people (Omaha, Nebraska, in this case) includes at least a half-dozen East Asian grocery stores, with goods from China and nearby countries, and at least two specializing in goods from India and Bangladesh. Another sells groceries from the Philippines and Vietnam. Main-line grocery stores now also sell extensive offerings from East Asia and Latin America. Omaha has three Spanish-language newspapers that carry advertising from groceries. Many of these goods are shipped from thousands of miles away, utilizing oil-based fuels.

Cruising the aisles of our local Hy-Vee (a regional grocery chain), within an hour I came up with a short list of long-distance imports. Some of these goods probably were air-shipped, but most reached Hy-Vee via surface, ships, trucks, and trains, which involves longer distances but is much more energy-efficient. These are only a few examples of what is available (the figures below are statute miles and kilometers to New York City from a major city in the place of origin; for surface miles via ocean transport, add 10 percent).

Alaska: salmon (Anchorage: 3,371 miles/5,425 kilometers)

Australia: winter nectarines (Sydney: 9,935/15,989)

Belgium: banana nectar (no kidding!) (Brussels: 3,600/5,840)

Chile: grapes and other winter fruits (Santiago: 5,107/8,219)

China: beer (Beijing: 6,843/11,012)

Costa Rica: pineapples and bananas (San Jose: 2,509/4,039)

Ecuador: cut flowers, tilapia (Quito: 2,834/4,560)

Germany: beer (Frankfurt: 3,858/6,208)

Honduras: shrimp (Tegucigalpa: 2,004/3,225)

Hong Kong: jasmine rice (8,059/12,968)

Italy: premium pasta (Rome: 4,283/6,891)

Kenya: tilapia, shrimp (Nairobi: 7,358/11,842)

Mexico: mangoes, papayas, watermelon, cantaloupe, winter strawberries (Mexico City: 2,090/3,363)

Netherlands: flower seeds (Amsterdam: 3,652/5,877)

Nicaragua: winter watermelon, shrimp (Managua: 2,310/3,719)

Peru: winter asparagus (Lima: 3,640/5,857)

Philippines: mangoes (Manila: 8,509/13,693)

Russia: caviar (Moscow: 4,668/7,511)

Spain: olives, artichoke hearts (Madrid: 3,591/5,779)

United Kingdom: Coleman's mustard (London: 3,463/5,772)

Venezuela: winter watermelon (Caracas: 2,149/3,459)

Within the United States, many fruits and vegetables are shipped nationwide from California (Los Angeles–New York: 2,451 miles/3,944 kilometers), at what would be international distances in many other countries. In summer, however, produce is more localized. The area around Omaha produces some of the best watermelon on Earth, but for only two to three months (August–October).

The fish counter of one Omaha supermarket *alone* could serve as a virtual United Nations of seafood. My wife obtained a chart from the fishmonger at our neighborhood Hy-Vee. At one time or another, the chain sells fish and shellfish of about eighty varieties (alphabetically, from Arctic char to whitefish) from thirty-five countries: the United States, Canada, Mexico, Costa Rica, Chile, Ecuador, Colombia, U.S. Virgin Islands, China, Honduras, Panama, Bahamas, Suriname, Belize, Guatemala, Venezuela, Nicaragua, Brazil, the Philippines, Namibia, Indonesia, Australia, New Zealand, India, Thailand, the United Kingdom, France, Peru, India, Vietnam, Bangladesh, Malaysia, Pakistan, Sri Lanka, and Brazil. The only types of fish obtained solely from the United States (usually many hundreds of miles from land-locked Omaha) are carp, Alaskan cod, oysters, white wild shrimp (the chain sells seventeen varieties of shrimp), catfish, and Idaho rainbow trout (Hy-Vee 2007).

## LOCALIZING THE FOOD CHAIN

Concern regarding global warming has helped stoke a movement toward locally grown food. Even as the supply lines of many grocery chains lengthen, more urban residents are raising food on vacant lots. By 2007, two thousand community gardens were operating in U.S. urban areas. In 2007, novelist Barbara Kingsolver's description of how her family spent a year eating local or homegrown food (*Animal, Vegetable, Miracle*, Harper-Collins) landed on the *New York Times* hard-cover best-seller list (at number 9 on August 12). The Sustainable Table website (www.sustainabletable.org), which addresses food and energy issues, tracks such things.

The shortest distance of all, of course, is from one's backyard (or roof-top) to the kitchen. I hear via moccasin telegraph that some New York City residents who have access to rooftops have been making a purposeful effort to localize their eating habits by raising all manner of vegetables—and even honey (with attendant bees) as well as the occasional chicken, for laying and eating.

More city dwellers are raising their own chickens. As with urban clothes-lines, however, city chickens frequently encounter legal barriers and irritated neighbors. Urban chickens have become the rationale for cottage industries. Internet pages have proliferated (see, for example, www.thecitychicken.com or www.backyardchickens.com). In Los Angeles, Phoenix, and Austin, residents have organized chicken-centric social events, and according to one report, "dozens of books—a whole new form of chick lit—on raising chickens, including Barbara Kilarski's *Keep Chickens! Tending Small Flocks in Cities, Suburbs and Other Small Spaces*, and related titles like *Anyone Can Build a Tub-Style Mechanical Chicken Plucker*, by Herrick Kimball" (Price 2007). The movement even has its own magazine, *Backyard Poultry*, founded in 2006.

Some cities ban in-city chickens, especially crowing roosters, but New York, Oakland, San Francisco, Houston, Chicago, Seattle, and Portland, Oregon, among others, have laws permitting them. A "poultry under-ground" in Madison, Wisconsin, agitated and won legal permission during 2005. Their story was told on Madcitychickens.com. In cities such as Boston, where all chickens are illegal, many people keep small coops anyway. "We catch them all the time," said a representative from Boston's animal control unit, who asked to remain anonymous because he was not author-ized to speak to the news media about poultry. "There's chickens all over the place" (Price 2007).

Some U.S. grocery chains that specialize in organic foods, such as Whole Foods Market, among others, make an effort to purchase locally grown fruits and vegetables. Neighborhood grocery chains also purchase locally in season some of the time in areas with extensive agricultural bases (corn in Nebraska and Iowa, for example). Even Wal-Mart, as part of its never-ending search for ways to save money, recently has been finding ways to localize its supply chain.

Omaha, Seattle, New York, Boston, and many other urban locales host farmers' markets in season. In Omaha, at least, some of the best-tasting produce in the world comes from small fruit and vegetables stands that pop up around the city every summer and fall. That Nicaraguan watermelon tastes like cardboard next to the Norfolk melons that we have been buying for years at roadside stands.

Some farmers near U.S. urban areas also offer "organic box" plans, by which they supply produce on contract. My wife and I tried such a thing one year and found that we received all sorts of things we didn't want. The food also was often low quality. The farmer appeared to be saving his best and most saleable produce for the open market where he didn't have a captive clientele. Thus, this was a good idea in theory that did not work well in practice—at least in our singular experience. Many states have organic food certification groups, such as the Maine Organic Farmers and Gardeners Association (MOFGA), formed in 1971. Another is the California Certified Organic Farmers (CCOF). For more information, see the National Sustainable Agriculture Information Service (http://attra.ncat. org/).

## THE GREEN RESTAURANT MOVEMENT

Interest in eco-friendly eateries has provoked the formation of a Green Restaurant Association (GRA) in the United States. The seventeen-year-old group had certified three hundred eating establishments by 2007. Certification is earned by buying locally, avoiding plastic and Styrofoam, making recycling a priority, using environmentally friendly cleaners, using energy-efficient appliances, and sending used frying oil to be used as biodiesel fuel, among other things. The GRA's publicity points out that the one million restaurants in the United States consume more electricity than any other type of retail business. The average restaurant turns out 50,000 pounds of garbage (most of it organic) and uses 300,000 gallons of water per year (Stukin 2007, 52).

Customers at American Flatbread, a pizzeria in the Washington, D.C., suburb of Ashburn, Virginia, drink beer that is brewed nearby. Leaves of their iced tea are grown and packaged on a local farm. The restaurant's owners acquire scallions, spinach, cheese, and corn within a few miles. On the restaurant's back wall, a hand-painted map of Loudoun County shows the farms and dairies that the restaurant uses to assemble its menu. Its wood-fueled oven is made of Virginia red clay (Lazo 2007, A-1). Several other restaurants in the same area have found that using local foods is tasty, popular, and profitable, suiting an area of mixed suburbs and farms that have been struggling to survive urban sprawl. Local farms supply local groceries (including burgeoning organic food markets) as well as restaurants with corn, melons, tomatoes, peppers, squash, eggplants, cucumbers, and even lamb. American Flatbread's frozen pizzas also are sold at local grocery stores.

At the Habana Outpost in Brooklyn, a solar-panel awning shades the outdoor dining area, providing the energy for the first solar-powered restaurant in New York City. A bicycle crank provides the power that blends fruit smoothies. "I wanted to make being environmentally responsible fun and friendly, without being didactic," said the eatery's owner, Sean Meenan (Stukin 2007, 51). Cell phones and computers are solar-powered. Toilets use rainwater, kitchen waste is composted, and utensils and drinking cups are made of cornstarch. Meenan, who also owns two other eco-eateries in New York City, also buys his organic food as close to home as possible.

During the growing season, the owners of McFoster's Natural Kind Café in Omaha grow 70 percent of the vegetables its customers eat in their own 16-acre chef's garden, and have been doing so since 1994. The local harvest not only provides fresh produce and reduces the carbon footprint of the restaurant's food but also allows it to use exotic varieties that can't withstand shipping and thus rarely show up in markets—purple dragon carrots, anyone? Foster also grows edible flowers that grace plates in the restaurant. A self-described "environmental rabble-rouser," Foster and his life partner Mary Helen McGranaghan (the restaurant is an amalgam of their names) maintain an organic farm, avoiding synthetic herbicides, pesticides, and fertilizers for compost and horse manure. "No chef or restaurant operator is going to turn down a lug of home-grown tomatoes for those hard, orange ones you get shipped from Argentina," he says (Aksamit 2007, 2-E).

## REFERENCES

Abboud, Leila. 2007. "How Denmark Paved Way to Energy Independence." *Wall Street Journal*, April 16, A-1, A-13.

Aksamit, Nicole. 2007. "Young McFoster Had a Farm." *Omaha World-Herald*, August 22, 1-E, 2-E.

Ball, Jeffrey. 2007a. "The Carbon Neutral Vacation." *Wall Street Journal*, July 30, P-1, P-4–5.

———. 2007b. "Climate Change's Cold Economics: Industry Efforts to Fight Global Warming Will Hit Consumers' Pockets." *Wall Street Journal*, February 15, A-12.

Bly, Laura. 2007. "How Green Is Your Valet and the Rest?" *USA Today*, July 12, D-1, D-2.

Bradsher, Keith. 2007. "Push to Fix Ozone Layer and Slow Global Warming." *New York Times*, March 15. http://www.nytimes.com/2007/03/15/business/worldbusiness/15warming.html.

Butler, Declan. 2007. "Super Savers: Meters to Manage the Future." *Nature* 445 (February 8): 586–88.

Deutsch, Claudia H. 2007b. "Trying to Connect the Dinner Plate to Climate Change." *New York Times*, August 29. http://www.nytimes.com/2007/08/29/business/media/29adco.html.

Eilperin, Juliet. 2007a. "Beyond Wind and Solar, a New Generation of Clean Energy." *Washington Post*, September 1, A-1. http://www.washingtonpost.com/wp-dyn/content/article/2007/08/31/AR2007083102054_pf.html.

Environment News Service. 2007a. "Australia Screws in Compact Fluorescent Lights Nationwide." February 21. http://www.ens-newswire.com/ens/feb2007/2007-02-21-01.asp.

———. 2007d. "British Columbia to Trim Greenhouse Gases, Go Carbon Neutral." February 14. http://www.ens-newswire.com/ens/feb2007/2007-02-14-02.asp.

———. 2007e. "Building Parks Can Help to Climate Proof Cities." June 12. http://www.ens-newswire.com/ens/jun2007/2007-06-12-04.asp.

———. 2007k. "College and University Presidents Pledge Climate Neutral Campuses." February 22. http://www.ens-newswire.com/ens/feb2007/2007-02-22-09.asp#anchor6.

———. 2007z. "University of California Adopts Green Purchasing, Climate Policies." April 4. http://www.ens-newswire.com/ens/apr2007/2007-04-04-09.asp#anchor5.

Fialka, John J. 2007b. "U.S. Plots New Climate Tactic." *Wall Street Journal*, September 7, A-8.

Frank, Robert. 2007. "Living Large While Being Green." Wealth Report. *Wall Street Journal*, August 24, W-2.

Gaarder, Nancy. 2007a. "Many Digging Deep for Cheaper Energy." *Omaha World-Herald*, May 29, A-1, A-2.

Harden, Blaine. 2006. "Tree-Planting Drive Seeks to Bring a New Urban Cool; Lower Energy Costs Touted as Benefit." *Washington Post*, September 4, A-1. http://www.washingtonpost.com/wp-dyn/content/article/2006/09/03/AR2006090300926_pf.html.

Hillman, Mayer, Tina Fawcett, and Sudhir Chella Rajan. 2007. *The Suicidal Planet: How to Prevent Global Climate Catastrophe*. New York: St. Martin's/Dunne.

Hughes, Kathleen. 2007. "To Fight Global Warming, Some Hang a Clothesline." *New York Times*, April 12. http://www.nytimes.com/2007/04/12/garden/12clothesline.html.

Hy-Vee. 2007. "Hy-Vee Seafood Country of Origin." Peony Park Hy-Vee, Omaha, Neb. September 27. Photocopy.

Jordan, Steve. 2007a. "Becoming Greener: Environmental Concerns, Lower Costs Push Drive for Energy-Efficient Buildings." *Omaha World-Herald*, July 22, D-1, D-2.

Knoblauch, Jessica A. 2007. "Have It Your (the Sustainable) Way." *EJ* (*Environmental Journalism*), Spring, 28–30, 46.

Lazo, Alejandro. 2007. "A Shorter Link between the Farm and Dinner Plate: Some Restaurants, Grocers Prefer Food Grown Locally." *Washington Post*, July 29, A-1. http://www.washingtonpost.com/wp-dyn/content/article/2007/07/28/AR2007072801255.html.

Lohr, Steve. 2007. "Energy Standards Needed, Report Says." *New York Times*, May 17. http://www.nytimes.com/2007/05/17/business/17energy.html.

MacLeod, Calum. 2007. "China Envisions Environmentally Friendly 'Eco-city.'" *USA Today*, February 16, 9-A.

Masters, Coco. 2007a. "End the Paper Chase." *Time*, March 27. http://www.time.com/time/printout/0,8816,1603633,00.html.

———. 2007c. "Kill the Lights at Quitting Time." *Time*, March 27. http://www.time.com/time/printout/0,8816,1603601,00.html.

———. 2007d. "Make Your Garden Grow." *Time*, March 27. http://www.time.com/time/printout/0,8816,1603649,00.html.

———. 2007e. "Rake in the Fall Colors." *Time*, March 27. http://www.time.com/time/printout/0,8816,1603631,00.html.

———. 2007f. "Shut Off Your Computer." *Time*, March 27. http://www.time.com/time/printout/0,8816,1603535,00.html.

Ouroussoff, Nicolai. 2007. "Why Are They Greener Than We Are?" *New York Times Sunday Magazine*, May 20. http://www.nytimes.com/2007/05/20/magazine/20europe-t.html.

Price, Catherine. 2007. "A Chicken on Every Plot, a Coop in Every Backyard." *New York Times*, September 19. http://www.nytimes.com/2007/09/19/dining/19yard.html.

Revkin, Andrew C., and Patrick Healy. 2007. "Coalition to Make Buildings Energy-Efficient." *New York Times*, May 17. http://www.nytimes.com/2007/05/17/us/17climate.html.

Rogers, Paul. 2007. "New American Home: A Showcase for Energy Efficiency, 'Green' Design." *Omaha World-Herald*, March 11, RE-1, RE-2.

Romm, Joseph J. 2007. *Hell and High Water: The Solution and the Politics—and What We Should Do*. New York: William Morrow.

Sanders, Robert. 2001. "Standby Appliances Suck Up Energy." *Cal Neighbors*, Spring. http://communityrelations.berkeley.edu/CalNeighbors/Spring2001/appliances.html.

Speth, James Gustave. 2004. *Red Sky at Morning: America and the Crisis of the Global Environment*. New Haven, Conn.: Yale University Press.

Steinman, David. 2007. *Safe Trip to Eden: 10 Steps to Save Planet Earth from Global Warming Meltdown*. New York: Thunder's Mouth Press.

Stukin, Stacie. 2007. "The Lean, Green Kitchen." *Vegetarian Times*, September, 51–54.

U.S. Green Building Council. 2008. "LEED Rating Systems." http://www.usgbc.org/DisplayPage.aspx?CMSPageID=222.

Walsh, Bryan. 2007b. "Think outside the Packaging." *Time*, April 9. http://www.time.com/time/printout/0,8816,1603638,00.html.

White, Martha C. 2007. "Enjoy Your Green Stay." *New York Times*, June 26. http://www.nytimes.com/2007/06/26/business/26green.html.

Will, George F. 2007. "Fuzzy Climate Math." *Washington Post*, April 12, A-27. http://www.washingtonpost.com/wp-dyn/content/article/2007/04/11/AR2007041102109_pf.html.

# CHAPTER 4

# Biofuels: Where the Money Is

The U.S. government has been doing what it does best: turning a planetary emergency into a treasure trove for businesses that have enough money to hire squadrons of lobbyists. The problems, in this case, are dependence on imported oil and rising atmospheric levels of greenhouse gases. The magic word, as of 2007, was "ethanol," that is, fuel from biomass—corn, usually, in the new "green" calculus of the federal government.

I noted in my browsings one late-summer day that during 2006 the United States for the first time had replaced Brazil as world champion of ethanol production, 5 billion gallons to 4.2 billion. Blessed by President George W. Bush in the waning years of what seemed like the longest lame-duck presidency on record, the ethanol boom was being driven by a 51-cent-a-gallon federal subsidy, as biofuel fever struck Capitol Hill. The ethanol gravy train received a gigantic push from Bush in his 2007 State of the Union message, during which he vowed to reduce U.S. consumption of imported oil, calling for a fivefold, government-mandated increase in ethanol production.

The idea of using carbohydrates as fuel is not new. At the dawn of the automotive age, Henry Ford believed that "ethyl alcohol is the fuel of the future" (Rohter 2006); Ford's first car was designed to run on it. Rudolf Diesel propelled his first engine with peanut oil. During World War II, ethanol was used to stretch supplies of gasoline. Never before, however, has corn ethanol been such a tax-assisted boondoggle.

Good or bad, right or wrong, the United States' national commitment to ethanol by 2007 included current tax credits, grants, and loan guarantees that could cost the federal treasury $140 billion in fifteen years. New proposals under consideration in Congress at the time could raise the eventual cost to $205 billion. The biggest single prospective expense was an extension of the 51-cent-a-gallon ethanol tax credit that had been scheduled to expire in 2010. The extension could cost an estimated $131 billion through 2022. The national commitment did *not* include abolition of the 54-cent-a-gallon tariff against imported ethanol such as Brazilian sugarcane-based biofuel, which yields eight times the energy of corn. That one scores no points in Lobby Alley on K Street, where environmental sanity still plays second fiddle to special domestic interests.

One bill under consideration at that time required that use of biofuels climb to 36 billion gallons by 2022, more than six times the capacity of the nation's 115 ethanol refineries. "There's almost a gold rush in this sector at the moment," said Philip R. Sharp, who, after serving in the House of Representatives for twenty years, in 2007 was a lobbyist as president of Resources for the Future (Mufson 2007e, D-1). Congressional projections of ethanol production contain a combination of ignorance, wish fulfillment, and downright political silliness. One sober scientific analysis estimates that to produce the amount of corn ethanol that Bush wants by 2017 would require half the world's land surface (that is, more than its total cultivated land) (Thomas 2007, 349).

In the midst of all this corn-fed hype, a report compiled by the Organization for Economic Cooperation and Development (a forum for developed nations) recommended during 2007 that the United States and other industrialized nations eliminate ethanol subsidies. "The overall environmental impacts of ethanol and biodiesel can very easily exceed those of petrol and mineral diesel," the report said (*New York Times* 2007).

The greening that matters here is the folding kind. Driving through Iowa along State Route 30 northeast of Omaha, to meet NASA climate scientist James Hansen in his hometown of Denison during the spring of 2007, my wife and I found ethanol plants popping up across the Midwest like mushrooms after a spring rain, pumping away in cheap aluminum casings next to corn silos and railroad tracks. We could nearly smell the subsidies. In addition to the ethanol tax credit, other incentives include a $1-a-gallon biodiesel tax credit, a subsidy for service stations that install E85 pumps, spending by the Agriculture Department on energy programs, and various other Department of Energy (DOE) grants and loan guarantees (Mufson and Morgan 2007, D-1).

Despite the hype, corn ethanol is a monumentally bad idea. The imbalance between the amount of energy in corn ethanol and what we are consuming as gasoline renders impossible the potential that one will replace the other unless most of us junk our cars and take up some *very* serious bicycling. Filling the gas tank of a sports-utility vehicle with ethanol *once* requires 450 pounds of corn, enough to feed a person for a year. While sugarcane ethanol produces eight times the energy required to produce it (and gasoline produces five times), corn ethanol, energy-wise, is a wash, at 1.3 times the energy required for manufacture. Replacing half the 2006 gasoline consumption in the United States with corn ethanol would require about seven times the land area that the United States has planted with corn (Goodell 2007, 50, 53).

Despite such problems, ethanol has become the new "other white meat" (i.e., pork). One proposal introduced in Congress during 2007 included aid for ethanol infrastructure (ethanol corrodes gasoline pipelines). Another proposal establishes a Strategic Ethanol Reserve for years when corn harvests are reduced by droughts, similar to the national reserve for petroleum. Then there are billions of dollars for research into cellulosic ethanol technologies, an estimated $10.8 billion by 2007. *Ka-ching!* (Mufson and Morgan 2007, D-1).

## DO BIOFUELS EMIT MORE GREENHOUSE GASES THAN FOSSIL FUELS?

By March 2008, with corn futures rising again to more than $5.50 a bushel, scientists evaluated ethanol's full environmental cost. When the full emissions costs of producing biofuels are calculated, most of them are environmentally more expensive, causing more greenhouse gases than fossil fuels, according to studies published in *Science* early in 2008. Growth of feedstock for many biofuels, from corn to sugar cane to palm oil, destroys natural ecosystems (most notably rain forest in the tropics and South American grasslands), releasing gases as they are burned and plowed. Destruction of these older, natural ecosystems also removes carbon sinks. In addition to the greenhouse gases caused by growing biofuels, additional emissions result from refining and transporting them.

"When you take this into account, most of the biofuel that people are using or planning to use would probably increase greenhouse gasses substantially," said Timothy Searchinger, lead author of one of the studies and a researcher in environment and economics at Princeton University. "Previously there's been an accounting error: land use change has been left out of prior analysis" (Rosenthal, 2008).

Clearance of grassland releases 93 times the amount of greenhouse gas that would be saved by the fuel made annually on that land, said Joseph Fargione, lead author of the second paper, and a scientist at the Nature Conservancy. "So for the next 93 years you're making climate change worse, just at the time when we need to be bringing down carbon emissions" (Rosenthal 2008).

Many U.S. farmers are growing corn year-round, which used to alternate with soybeans. More soybeans are being raised on newly cleared rain forest land in Brazil. Joseph Fargione and colleagues wrote:

> Biofuels are a potential low-carbon energy source, but whether biofuels offer carbon savings depends on how they are produced. Converting rainforests, peatlands, savannas, or grasslands to produce food crop-based biofuels in Brazil, Southeast Asia, and the United States creates a "biofuel carbon debt" by releasing 17 to 420 times more $CO_2$ than the annual greenhouse gas (GHG) reductions that these biofuels would provide by displacing fossil fuels. In contrast, biofuels made from waste biomass or from biomass grown on degraded and abandoned agricultural lands planted with perennials incur little or no carbon debt and can offer immediate and sustained GHG advantages. (Fargione et al. 2008, 1235).

A second scientific team confirmed these findings in the same edition of
*Science*. Timothy Searchinger and colleagues wrote:

> Most prior studies have found that substituting biofuels for gasoline will
> reduce greenhouse gases because biofuels sequester carbon through the
> growth of the feedstock. These analyses have failed to count the carbon
> emissions that occur as farmers worldwide respond to higher prices and con-
> vert forest and grassland to new cropland to replace the grain (or cropland)
> diverted to biofuels. By using a worldwide agricultural model to estimate
> emissions from land-use change, we found that corn-based ethanol, instead
> of producing a 20 percent savings, nearly doubles greenhouse emissions over
> 30 years and increases greenhouse gases for 167 years. Biofuels from switch-
> grass, if grown on U.S. corn lands, increase emissions by 50 percent. This
> result raises concerns about large biofuel mandates and highlights the value
> of using waste products. (Searchinger et al. 2008, 12386).

Some of the carbon footprint of biofuels may be mitigated by conservation tilling, which minimizes disturbance of the soil to retain carbon dioxide, and, at the same time, leaves at least 30 percent of the surface covered
with crop residues. Use of such methods can reduce carbon dioxide emissions from the soil by as much as 50 to 60 percent. Crop residue also can
provide feedstock for cellulosic ethanol (Richards 2008, 7-B).

## U.S. ETHANOL CAPACITY SURGES

Regardless of its problems (stoked by the governmental gravy train),
ethanol production capacity in the United States grew by 400 percent
between 2000 and 2006, to 5.6 billion gallons a year, or 20 percent of
the country's corn crop, compared with less than 5 percent in 2000
(Odling-Smee 2007, 483). In August 2006, U.S. refineries were producing 27 percent more ethanol than a year earlier, and forty-eight distilleries
were under construction (Mufson 2006, D-1). In two years and nine
months (from January 2005 to September 2007), forty-eight new ethanol
distilleries opened in the United States (raising the count from 81 to
129), most of them in Midwest corn country.

By 2007, local newspapers in the Midwest (one was our *Omaha World-
Herald*) were filled with news describing how farmers were planning to
plant corn fencepost-to-fencepost to cash in on the bonanza. By October
2007, U.S. farmers were harvesting a record 13.3 million bushels of corn,
busting the old record of 11.8 million in 2004. Capacity was increasing so
quickly that a glut developed, with wholesale ethanol selling for 60 cents a
gallon less in Iowa than New York City (Krauss 2007). As the price of
corn remained high and their product's price dropped, ethanol distillers
faced a shakeout late in 2007, at least for a time.

A doubling of ethanol production (a feasible figure within a few years if
plants now being planned are built) would put 40 percent of the U.S.
corn crop into ethanol. Congress in 2005 enacted a law requiring ethanol
production of 7.5 billion gallons by 2012. In 2007, several senators from
the Midwest introduced a bill called the Biofuels Security Act that aimed

to increase that figure by about 90 percent. It didn't pass—probably a good thing because no one seemed to have done the math. The law would have required conversion to ethanol of much more corn than the country has ever produced, leaving nothing to eat unless corn production increased enormously. By March 2007, the price of corn was $4.10 a bushel, up from less than $2 during 2005—more than 100 percent in a year. The price later ebbed to about $3.50.

Ethanol's production also was being spurred by its use as a substitute for methyl-$t$-butyl ether (MTBE), an additive used to lower emissions in gasoline that has been shown to be a health risk. Thus, for all these reasons, major investors, including Archer Daniels Midland (ADM), have jumped on ethanol as their next big profit maker. By 2006, ADM was manufacturing about 20 percent of U.S. ethanol.

To prove that plant-based fuels have all the machismo of gasoline, on May 27, 2007, all cars competing in the Indianapolis 500-mile auto race, which attain speeds of 220 miles an hour, for the first time ran on corn-based ethanol. Racing champion Bobby Rahal had announced the change earlier in May, calling it "a tribute to the spirit of American ingenuity and innovation." "The use of 100 percent fuel-grade ethanol makes the Indy Car Series the first in motor sports anywhere in the world to embrace a renewable and environmentally friendly fuel source," he said at the National Press Club on May 4 (Environment News Service 2007u). In 2006, the race, in which some cars run at 675 horsepower, had been run on an ethanol-methanol mix. Pure methanol had been the preferred fuel at the race for four decades. After five auto races in 2007 with no signs of inferior engine performance, drivers have accepted ethanol. John Griffin, vice president of public relations for the Indy Car Series said, "At the end of the day, they see that we are doing something good for the environment without losing anything" (Environment News Service 2007u).

Indy 500 drivers have an advantage over the average motorist using nearly pure ethanol: a ready source of fuel at trackside. Driving across the United States in search of E85 is still something of a challenge. To begin with, only about 1,000 of 179,000 gas pumps at fueling stations in the United States offered it, as of late 2007 (Krauss 2007). Even with an electronic locator provided by the Department of Energy (www.eere.energy.gov/afdc), one might have to drive 50 to 100 miles off a straight route just to find a gas station with an E85 pump (Vella 2007, D-2). Outside the Midwest, where E85 pumps are more widespread than elsewhere, searching for it can be a real chore for drivers of the five million cars in the United States that have been equipped to use the fuel.

Choice of vehicles is limited, as well. The DOE provides a list of thirty-one E85 compatible vehicles (see www.fueleconomy.gov), mostly heavy SUVs, pickup trucks, large vans, or passenger cars such as the Ford Crown Victoria or Saturn Relay. Fuel-efficient vehicles are rare. Since ethanol is less fuel-efficient than gasoline, biofueled vehicles might get as little as 9 miles per gallon (mpg) in city driving. A Chevy Suburban (typical of the available models) gets 14 to 20 mpg on gasoline and 10 to 15 mpg on E85 (Vella 2007, D-2).

## JUST HOW GREEN IS IT?

The ethanol gravy train has been very good for farmers where I live. The land under their farms is even rising in value, and the ethanol juggernaut is feeding the state treasury that pays part of my salary. Prices for farmland across the United States soared 13.7 percent between 2006 and 2007. Thus, I regret raining on the ethanol parade. However, pitching ethanol as an environmental solution to fossil fuels is, as we say around here, "putting lipstick on the pig."

The ethanol parade has been propelled by an assumption that it's an environmentally friendly ride. Just how "green" is it? Corn, processed into fuel, is still carbon and, burned to propel cars, it is still like a fossil fuel. Most corn ethanol emits only 10 to 20 percent less greenhouse gases per mile driven than gasoline. Even that savings isn't what it seems. Corn must be grown on factory farms, a very energy-intensive business itself. Under some circumstances (if a biomass field replaced a forest, for example), this type of fuel might actually produce a net increase in relative emissions of greenhouse gases.

In the meantime, the price of corn has been rising rapidly worldwide, feeding inflation of food prices. You'd be surprised just how widespread the use of corn has become in our food supply, aside from the obvious products such as corn flakes. Meat animals eat it, so it affects the price of chicken (meat and eggs), beef, and pork. The manufacture of soda pop consumes huge amounts of corn syrup, so its price also rises.

In addition to raising prices, increasing the amount of land planted in corn for ethanol production (or any other reason) increases water pollution. Corn requires more fertilizer than soybeans, hay, or many other crops. The added load of nitrogen and phosphorus feeds unnatural algae blooms in local watercourses. The algae consume oxygen that would otherwise have been available to fish and other aquatic creatures, creating or expanding oxygen-starved dead zones in our watercourses and into the oceans (Fahrenthold 2007, B-1).

The rising number of ethanol plants in the Great Plains region may drain billions of gallons of water each year from the Ogallala Aquifer, which already is depleted, Environmental Defense asserts. A report authored by Martha Roberts and Theodore Toombs describes the aquifer as one of the world's largest, a vast, shallow underground pool beneath portions of Colorado, Kansas, Nebraska, New Mexico, Nebraska, Oklahoma, South Dakota, Texas, and Wyoming. The Ogallala Aquifer already supports agriculture in the area, where the water table is declining as rates of groundwater pumping regularly exceed replacement. A gallon of ethanol requires four gallons of water to produce. "This dramatic expansion of ethanol production has substantial implications for already strained water and grassland resources in the Ogallala Aquifer region," Roberts and Toombs assert (Environment News Service 2007m).

Other concerns include truck traffic in rural areas, and air pollution with a sticky-sweet smell that resembles that of a barroom floor after a busy Saturday night (Barrett 2007, A-1). Residents in Webster County,

Missouri, sued to stop construction of an ethanol plant on grounds that it would use more water than the county's 33,000 residents combined.

Research by David Pimentel, a professor at Cornell University, and Tad Patzek, a professor of chemical engineering at the University of California at Berkeley, indicated that with current processing methods, more fossil energy is used to produce the energy equivalent in biofuel than it contains.

Even when research includes the energy and labor necessary to build processing plants and farm machinery, which usually are not included, it has not included the cost of waste treatment or the environmental impact of intensive bio-energy crops, such as the loss of soils and environmental pollution due to the use of fertilizers or pesticides. Patzek's calculations indicate that biofuels are not carbon neutral. The rush to subsidize ethanol ignores processing that depends on fossil fuels. Some of the 114 ethanol plants in the United States use natural gas and even coal to run their processors. Ethanol also must be transported by truck to points of sale because it corrodes metal gas pipelines (Patzek and Pimentel 2005, 327–64; Patzek 2006, 1747). Thus, more fossil fuels are required to transport ethanol than fuel that moves in pipelines.

## DOING THE MATH: AN ETHANOL CRITIQUE

One major problem with biofuels is the amount of land required to grow the basic feedstock. Writing in Britain, author George Monbiot calculated that using the United Kingdom's most productive oil crop (rapeseed) to replace petroleum in British transport at present-day consumption levels would require four to five times the nation's entire arable land base:

> In order to move our [British] cars and buses with biodiesel, we would require 25.9 million hectares. There are 5.7 million hectares in the United Kingdom. If this were to happen all over Europe, the consequences on food supply would be catastrophic: enough to tip the scales from being excess producers to becoming net losers. If, as some environmentalists claim, this were to be done on a world scale, most of the arable surface of the planet would have to be given over to producing food for cars, not for people. This outlook would seem, at a first glance, to be ridiculous. If the demand for food could not be covered, wouldn't the market ensure that crops be used to feed people instead of cars? Nothing is sure about this. The market responds to money, not to needs. (Monbiot 2006a, 158)

In addition, rapeseed depletes the soil (thus its name).

Researchers at the University of Minnesota have estimated that converting the entire U.S. corn crop to ethanol would replace only one-eighth of U.S. gasoline consumption (Krugman 2007, A-23). In addition, corn must be grown and transported, after which ethanol must be manufactured. Replacing a gallon of gasoline with a gallon of ethanol does not save a gallon of gasoline because most of the energy that goes into corn comes from fossil fuels. The real savings is more like a quarter of a gallon— so make that 3 percent savings of gasoline for the entire U.S. corn crop (Krugman 2007, A-23). Then, what would we *eat*?

## PROTESTS OVER RISING CORN PRICES

Once ethanol started pinching corn prices, senators' and representatives' phones lit up with protests from chicken farmers, pork producers, and representatives of the dairy industry, all of whom found feed prices rising with corn futures. Corn prices doubled to $4 a bushel, then eased to $3.40 as farmers in the Midwest increased their plantings 20 percent for the 2007 growing season. The price of popcorn rose 40 percent between 2006 and 2007. The National Turkey Federation estimated that feed costs rose $600 million a year (Strassel 2007, A-16). Some pork producers switched from corn to anything else they could find in bulk—French fries, tater tots, hash browns, cheese curls, candy bars, yogurt-covered raisins, dried fruit, and past-pull-date breakfast cereal mixed with chocolate powder. Sometimes farmers use burnt cookies, breakfast cereal with too much sugar for human consumption, noodles that had fallen off assembly lines while being packed, or trail mix sprinkled with cardboard (Etter 2007b, A-1, A-14). Ethanol is not a total debit for farmers, however. The manufacture of biodiesel ethanol from corn also produces as a byproduct distiller's grain that can be fed to animals.

Meanwhile, the National Cattlemen's Beef Association called for an end to tax subsidies of corn ethanol and an end to tariffs on foreign ethanol (such as Brazilian sugar-based fuel). Tyson Foods, the world's largest chicken producer and the second biggest employer in Arkansas (which also contains Wal-Mart's headquarters, in Bentonville), warned that rising corn prices would hurt its earnings. Coca-Cola, one among many large users of corn syrup in its soft drinks, hit the lobbying trail in Washington.

Ducks Unlimited complained to senators that the ethanol boom was causing many farmers to plant corn on formerly idle land that had been serving as duck habitat (Strassel 2007, A-16). Ducks aside, examined purely from a climatic perspective, radically increasing the amount of land planted in corn (or other crops) for ethanol also will release carbon dioxide that is bound in 35 million acres of marginal farmland now being held fallow for conservation and wildlife (Bourne 2007, 41).

Cal Dooley, president and chief executive officer of the Grocery Manufacturers' Association, criticized Bush's ethanol plan:

> In addition to its inflationary impact, there are many unintended, but nonetheless important, consequences of an ambitious corn-ethanol strategy. ... A 35 billion gallon ethanol mandate will require a substantial increase in the use of fossil fuels for corn and ethanol processing and transportation, as well as an additional 15 million acres devoted to corn crops, which will encroach on agriculturally marginal and environmentally sensitive land. (Environment News Service 2007f)

To meet this mandate, the United States would have to cut its corn exports to ensure an adequate supply of corn for food and fuel, Dooley said. "Such a reduction will result in a decrease in the amount of food available overseas, which in turn will have a negative effect on world hunger" (Environment News Service 2007f).

As early as January 31, 2007, rising prices for corn ignited demonstrations by 75,000 people in Mexico City, where the price of tortillas hit record highs as President Bush touted corn ethanol in his State of the Union message. Rising costs of farm goods, fed partially by demand for biofuels (corn, sugarcane, and palm oil, among others), have been pushing up food prices globally. This rise in prices was straining the food budgets among many poorer people in China, India, and other nations. Meanwhile, by late 2007, a third of the corn grown in Nebraska, the Cornhusker State, was going into ethanol, most of it into the state's sixteen distilleries. Corn was being diverted for the same reason Willie Horton said he robbed banks: "That's where the money is!" Fifty more ethanol plants were in planning stages across Nebraska (Bourne 2007, 42).

The trend was global, with grain stocks worldwide at a thirty-year low during 2007. China had only a two- to three-month grain supply in storage at that time. In Hungary, food-price inflation was running at 13 percent by March 2007, compared to 3 percent in 2005; in China, food prices were rising at 6 percent in 2007, versus 2 percent a year earlier. In the United States, food-price inflation was 3.1 percent annualized in February 2007, up from 2.1 percent in 2005. Food prices rose 15 percent a year in Turkey during 2007 (Barta 2007a, A-9). Food prices shot up even faster in 2008.

In Germany, the price of beer rose as farmers abandoned barley to grow crops that could be sold as feedstock for ethanol, such as corn. The barley crop fell 5.5 percent between 2005 and 2006, and prices rose. Prices at the 2007 Oktoberfest in Munich posted a 5.5 percent increase, raising a one-liter mug to the equivalent of $10.76. "Beer prices are a very emotional issue in Germany," said Helmut Erdmann, director of the Ayinger Brewery in the hills of Bavaria. "People expect it to be as cheap as other basic staples like eggs, bread, and milk." (Associated Press 2007d, A-18). Germany also is a leading manufacturer of biodiesel fuel from canola, another crop that has been competing with barley. In the United States, biodiesel fuel has been manufactured from soybeans, but output has been limited by low yield and high prices (almost $7 per gallon at 2007 prices), relative to fossil fuels. Biodiesel, however, does produce 60 to 70 percent less greenhouse-gas emissions compared to similar fuel derived from fossil fuels.

## ETHANOL FROM SUGARCANE: BRAZIL'S ENERGY CROP

Ethanol can be distilled from many natural substances, but sugarcane offers advantages over others such as corn. For each unit of energy expended to turn cane into ethanol, 8.3 times as much energy is created, compared with a maximum of 1.3 times for corn, according to scientists at the Brazilian Center for Sugarcane Technology. Sugarcane ethanol also generates 55 to 90 percent less carbon dioxide per unit of energy than corn. "There's no reason why we shouldn't be able to improve that ratio to 10 to 1," said Suani Teixeira Coelho, director of the National Center for Biomass at the University of São Paulo. "It's no miracle. Our energy balance is so favorable not just because we have high yields, but also because we don't use any fossil fuels to process the cane, which is not the case with corn" (Rohter 2006).

Using ethanol from sugarcane, Brazil became energy self-sufficient in 2006, even as demand for fuel grew. Brazil's full-court press on ethanol was three decades old by that time. During the 1970s, Brazil's government began developing the ethanol industry by subsidizing the sugarcane industry and requiring its use in government vehicles. By the late 1990s, the subsidies were phased out as the cost of producing ethanol dropped to 80 cents a gallon, less than the $1.50 worldwide average for producing gasoline at that time (Luhnow and Samor 2006, A-1, A-8).

Use of ethanol in Brazil accelerated after 2003 following introduction of "flex-fuel" engines, designed to run on ethanol, gasoline, or any mixture of the two. Gasoline sold in Brazil contains about 25 percent alcohol. By 2006, more than 70 percent of the 1.1 million automobiles sold in Brazil had flex-fuel engines (Rohter 2006). By 2007, sugarcane-based ethanol was supplying almost 40 percent of the energy used for ground transport in Brazil. Three-quarters of new cars sold in Brazil were equipped to use 85 percent ethanol, compared to 10 percent in the United States, and most fueling stations offered E85 fuel, compared to 1 percent in the United States.

If government was logical, the United States would be eating and exporting food from corn, and importing sugarcane-based ethanol from Brazil. We do, after all, import nearly everything else we consume. Government is driven by special interests, however, not rationality, and politicians from corn-growing states in the United States have obtained a stiff protective tariff on Brazilian sugar-derived alcohol. A 54-cent-a-gallon tariff on the imports plus a 51-cent-per-gallon subsidy on inferior domestic corn-based fuel equals a $1.05 tax advantage. Brazil exported $600 million worth of ethanol in 2005, but almost none of it came to the United States. The United States does grow some sugarcane along the Gulf of Mexico coast (and in parts of southeastern Florida), but government price supports make it too expensive to compete with other ethanol feedstocks.

Brazil's government taxes ethanol at 9 cents per gallon, compared to 42 cents for gasoline. All gasoline is legally required to contain at least 10 percent ethanol. Researchers in Brazil have decoded the genetics of sugarcane and used the knowledge to breed varieties with higher sugar content. Brazil has increased the per-acre productivity of sugarcane threefold since 1975 (Luhnow and Samor 2006, A-1, A-8).

"Renewable fuel has been a fantastic solution for us," said Brazil's minister of agriculture, Roberto Rodrigues. "And it offers a way out of the fossil fuel trap for others as well" (Rohter 2006). Sugarcane agriculture has its problems, however. The work is hot, dirty, and dangerous for the workers who harvest the cane by hand. To make fields easier to harvest (and kill snakes), fields are usually burned, adding carbon dioxide, methane, nitrous oxides, and soot to the atmosphere.

Over time, Brazil has made the manufacture of ethanol more efficient. In the past, for example, the residue remaining after cane stalks were compressed to squeeze out the juice was discarded. Today, Brazilian sugar mills use that residue to generate electricity to process cane into ethanol. Other byproducts fertilize the fields where cane is planted. Some mills are now producing so much electricity that they sell their excess to the national grid.

If sugar is such a good source of ethanol, why not use sugar beets, which the United States produces in abundance? The processing equipment for sugar beets and sugarcane is somewhat similar. The problem with sugar beets lies with their harvest cycle in temperate regions. While sugarcane is grown nearly year-round (and thus could supply a processing plant almost all the time), sugar beets are grown on an annual cycle and harvested in the fall. Thus, according to Kenneth P. Vogel, a professor in the University of Nebraska at Lincoln's Agronomy Department, a processing plant that costs hundreds of millions of dollars would run only for a few months a year. This hurdle could perhaps be overcome with technology (not yet designed) to switch from sugar beets to corn and other types of ethanol sources. Instead, ethanol interests on the Great Plains are looking at switchgrass (which is handled like hay), corn stover, and other plant biomass. The manufacturing technology for these needs to be developed as well (*Omaha World-Herald* 2007, 6-B).

## ENVIRONMENTAL COSTS OF PALM OIL

While listening to the biofuel chorus, we also should remember that anything that obtains fuel from plant mass is not *ipso facto* good for the environment. The source of biofuels is important. Sugarcane, soy-based, or other biofuels grown on former rainforest land in the Amazon Valley or Indonesia is hardly environmentally conscious. When the subject is palm oil, which in 2007 replaced soy as the world's most-consumed vegetable oil, the victim is usually tropical rain forest and the animals whose habitats are being turned into factory farms. In Sumatra, Indonesia, less than 10 percent of mammals and birds that lived in the rain forests survived once they had become palm plantations (Stone 2007, 1491).

Palm oil is used in products as diverse as ice cream and soap, as well as for biofuel. Such plantations have an ecological and greenhouse cost, however: rainforest land is being cleared with fire, which adds carbon dioxide and methane to the atmosphere while ruining the habitat of orangutans and other increasingly scarce animals (Kennedy 2007, 515).

Palm oil's high energy efficiency per unit makes it an excellent biodiesel fuel, and trucks can run entirely on it, if necessary (although most run on a mixture of palm oil and petroleum-based products). The price of palm oil, which reached $700 per crude ton in 2007, may reduce its use as a fuel because it has become too expensive to compete with other sources. Large tracts of land have been converted to palm-oil production in Indonesia, with plans to double production in perhaps a decade, destroying large amounts of animal habitat. Global palm-oil production in 2006 reached 37 million tons, 85 percent from Indonesia and Malaysia. The yield, averaging 6.9 tons per acre, is seven times that of soybean oil, according to the United Nations Food and Agriculture Organization (Stone 2007, 1491).

## FOOD, FUEL, OR FORESTS?

The World Rainforest Movement, based in Uruguay, asserts that monoculture factory farming for food and fuel is destroying forests around the world. Soybean plantations in Argentina are gradually displacing the quebracho forests in the Chaco, while in Paraguay they are replacing the Pantanal, the Mata Atlantica, and the Chaco, and in Brazil, the Pantanal, the Mata Atlantica, the Cerrado, and the Caatinga. Between 1990 and 2002, the planted area of oil palms, a source of feedstock for ethanol manufacture, increased by 43 percent worldwide, mostly in Indonesia and Malaysia. Between 1985 and 2000, oil-palm plantations have been responsible for 87 percent of deforestation in Malaysia, with plans to replace another 15 million acres of forest.

In Sumatra and Borneo, roughly 10 million acres of forest has been converted to oil-palm plantations. In Indonesia, thousands of indigenous people have been evicted from their lands. Deforestation often spreads via fire. The entire region is becoming a gigantic vegetable oil field. In Uganda, the destruction of tropical forests and indigenous forestlands has begun in order to produce palm oil and sugar.

Slashing and burning of forests to make way for plantations of oil palm releases enormous carbon reserves. Thus, the road to ethanol is hardly carbon neutral. In marshy forests where there is peat, once the trees are cut, the plantations parch the soil. When the peat dries, it oxidizes and releases even more carbon dioxide than the trees.

## CRITIQUE OF CORN AS FUEL FROM A CLIMATIC POINT OF VIEW

President George W. Bush's enthusiasm for ethanol was weak on climate security. "I am disappointed," said Sen. Jeff Bingaman (D-N.Mex.), chairman of the Senate Energy Committee. He said Bush was "completely silent" on energy efficiency and reduction of carbon dioxide from electric power plants, which contribute 40 percent of carbon dioxide emissions. "If this was a real effort to solve the climate problem, it would include large stationary sources and utilities," said Eileen Claussen, president of the Pew Center on Global Climate Change, a nonpartisan policy research group. Claussen and other environmental leaders also criticized the absence of any proposal for a mandatory cap on emissions of heat-trapping gases (Andrews and Barringer 2007).

Environmentally, ethanol is a nonstarter, according to James E. Hansen, director of the NASA Goddard Institute for Space Studies and one of the leading climate scientists in the United States. According to Hansen:

> A proposed national plan for 20 percent ethanol in vehicle fuels, envisaged to be derived in large part from corn, does more harm to the planet than good. It would do little to reduce $CO_2$ emissions. It would degrade retention of carbon in soils and forests, and it would strike hard at the world's poor through increased food prices. There are a variety of ways that renewable or other $CO_2$-free energies may eventually power vehicles. Governments should not dictate the nature of those solutions. Biofuels are likely to play a major part in our energy future. As a native Iowan, I like to imagine that the Midwest will come to the rescue of compatriots threatened by rising seas. Native grasses appropriately cultivated, perhaps with improved varieties, can draw down atmospheric $CO_2$. The prairies from Texas to North Dakota may contribute, if we get on with solving the climate problem before super-drought spreads from the west to the prairies. If we act soon, we can keep the prairies as productive land. (Hansen, April 12, 2007).

## CELLULOSIC ETHANOL

For thirty years, microbiology professor Susan B. Leschine, who teaches at the University of Massachusetts at Amherst, has been researching the "Q microbe" that has an ability to break down leaves and plant fibers into ethanol. Suddenly, she has found venture capitalists talking up the idea, and so she is now chief scientist at SunEthanol, a new firm with about a dozen employees in 2007. The firm has attracted an equity investment from VeraSun Energy, a large producer of corn ethanol, which is betting that these microbes will supply a technological breakthrough necessary to the manufacture of cellulosic ethanol from switchgrass, wood chips, and other plant fibers. SunEthanol is one of many companies that are searching for ways to make cellulosic ethanol commercially viable, lured by $385 million of DOE grants (Mufson 2007c, D-1).

Some biomass fuels have a net energy balance ranging from 2:1 to 6:1. One problem with such fuels is that biomass is not very high-energy for a given weight (it stores little solar energy compared to sugarcane). Straw and field waste just doesn't pack much punch as fuel. Thus, immense amounts of land would be required to grow enough of it to make a significant dent in present-day gasoline consumption. On the other hand, utilizing all of a corn plant for fuel (instead of just its kernels) would double its yield per acre as energy. As early as 2007, DOE was commissioning a few existing corn-ethanol plants to try technology that uses the entire corn stalk. One such grant, for as much as $80 million, will enable Poet, an ethanol producer based in Sioux Falls, South Dakota, to open a cellulosic ethanol plant by 2011. The proposed facility will have a capacity of 125 million gallons a year, 25 percent from formerly wasted corn cobs and fiber (Associated Press 2007a, D-1).

**Figure 4.1. Truck unloading wood chips that will fuel the Tracy Biomass Plant, Tracy, California.**
Credit: National Renewable Energy Laboratory, Photographic Information Exchange. U.S. Department of Energy.

Widespread, inexpensive use of cellulosic ethanol will require thus-far unachieved technological advances. It also may require up to 40 million additional acres devoted to growing plant material, as well as a sprawling new infrastructure for transforming the feedstock into fuel (Andrews and Barringer 2007). Cellulosic ethanol has the potential of being eight times as energy-efficient as corn-based ethanol because it needn't be converted into sugar before it is manufactured into fuel.

"The challenge," wrote Donald Kennedy, editor of *Science*, "is biochemical. Plant lignins occlude the cellulose cell walls; they must be removed, and then the enzymology of cellulose conversion needs to be worked out. The technology is complex. No commercial reactor has yet been built, although six are funded" (2007, 515). BP (formerly British Petroleum) and the universities of California and Illinois have undertaken a $500 million project to cross this scientific barrier. If the technological problems can be solved, switchgrass could produce twice as much ethanol per acre as sugarcane. One solution may be genetic modification of microbes from the guts of termites, a "natural cellulosic energy factory" (Bourne 2007, 54).

The largest cellulosic ethanol plant in 2007 was a demonstration-scale facility in Canada built by the Ontario-based Iogen Corporation, which produces about a million gallons a year. A commercial-scale factory would require annual capacity of at least 40 million gallons, according to Robert Dineen, president of the Renewable Fuels Association, which represents ethanol producers (Andrews 2007a).

During 2007, Abengoa Bioenergy, a Spanish energy company with head offices in Seville, selected the town of Hugoton in southwestern Kansas as the site of the first U.S. cellulosic ethanol plant, a $400 million project that will convert 700 tons a day of corn stover (stalks), wheat straw, milo stubble, switchgrass, and other biomass into fuel. The plant will have a capacity of 30 million gallons a year; it also will include an 85-million-gallon-a-year corn ethanol plant. The Hugoton project will be partly funded by $76 million from DOE.

In February 2007, DOE awarded up to $385 million to six companies, including Abengoa, for first-generation ethanol plants in Florida, Georgia, Iowa, Idaho, California, and Kansas. "These biorefineries will play a critical role in helping to bring cellulosic ethanol to market, and teaching us how we can produce it in a more cost-effective manner," said Energy Secretary Samuel Bodman, awarding the grants. "Ultimately, success in producing inexpensive cellulosic ethanol could be a key to eliminating our nation's addiction to oil" (Environment News Service 2007v).

## JATROPHA—AN ALTERNATIVE TO CORN ETHANOL?

If corn is such a rotten energy source, what's left, other than sugarcane, which doesn't do well in a temperate climate? The seeds of the jatropha fruit will grow almost anywhere—deserts, trash dumps, rock piles—and they can be manufactured into biodiesel fuel. Before it became an energy crop in India, jatropha was dismissed as a weed. Like many weeds, it needs little water or fertilizer, and no one eats it. The plant was imported to India by Portuguese traders. Its oil has been used there by native peoples for many centuries to light lamps.

India has been taking a close look at jatropha as an energy crop on land that is too dry or infertile to grow other crops. An energy industry centered on jatropha has been eyed by BP, which has joined with the British firm D1 Oils in a $90 million joint partnership to develop biodiesel fuel from jatropha. Australia-based Mission Biofuels had invested $80 million in the same idea by 2007 and had 80,000 acres under cultivation in India, aiming at 250,000. India's state railway uses the fuel to power some of its locomotives and has planted the bushes along its tracks. The estimated cost of producing a barrel of biodiesel from jatropha is $43, half that of corn and a third the cost for rapeseed. This estimate is based on a small amount of experience, however, and may vary widely (Barta 2007b, A-1, A-12; Fairless 2007, 652).

## BIOGAS FROM WASTES AND LANDFILLS

Cattle manure is prosaic (some cities burn organic garbage and extract methane; some pig farms do the same from manure). As of 2007, however, biogas was the biggest source of alternative fuel in the United States. Companies use biogas or biomass generation as part of their industrial processes. Weyerhaeuser, for example, generates electricity with wood waste combined with byproducts from pulp mills that once were discarded as useless. Such sources have been generating power for 5 to 10 cents per kilowatt-hour.

Ingenuity is as ingenuity does—in the case of energy innovation, finding resources in what used to be waste. Witness a BMW manufacturing plant in Spartanburg, South Carolina, that by 2007 was getting more than half its energy from a nearby landfill, using formerly wasted methane, meanwhile also sparing the atmosphere from several thousands of tons of greenhouse gases a year.

Sen. Ben Nelson (D-Neb.) during May 2007 introduced legislation creating federal tax credits, loans, and loan guarantees for biogas, methane fuel that can substitute for natural gas and is created from animal waste—notably Nebraska's more than six million cattle (four times the state's human population). This is an interesting application of Barry Commoner's idea, now several decades old, that a pollutant is usually a resource that is out of place. A ConocoPhillips refinery in Texas uses beef, pork, and chicken fat from a Tyson Foods facility. This product, called "renewable diesel," started production during 2007 (Ball 2007c, A-8).

E3 Biofuels, a biogas plant in Mead, Nebraska, used manure from thirty thousand cattle on an adjoining feedlot to provide fuel for an on-site ethanol plant. The plant was touted as the first "closed-loop" ethanol facility in the world, utilizing its own fuel to reduce its in-house carbon footprint to nearly zero, meanwhile producing 25 million gallons of ethanol a year for sale. The plant also uses as fuel methane emissions that would have gone into the atmosphere. With all of these strategies, the Mead plant produced more than fifteen times more fuel than a gasoline refinery or a corn-ethanol plant—"a revolutionary step forward," according to Dennis Langley, E3's chairman (Hord 2007b, D-1). By mid-2007, E3 had plans to double the Mead facility's capacity and to open fifteen other plants elsewhere in the United States. Five months later, however, the plant's owners filed for bankruptcy protection, citing mechanical breakdowns caused by faulty construction and financial difficulties (Hord 2007a, D-1).

## BIOGAS IN SWEDEN AND DENMARK

Sweden and Norway have the highest liquor taxes in the world, provoking large-scale smuggling from Denmark. Until recently, contraband seized by gold-and-blue-capped Tullverket (Swedish Customs) at Malmö (across Öresund Sound from Copenhagen) was poured down the drain. These days, however, a million illicit bottles a year are trucked to a sparkling new high-tech plant in Linköping (about 80 miles south-southwest of

Stockholm) that manufactures biogas fuel. Every busted booze smuggler has been drafted into Sweden's war against oil dependence and greenhouse gases. Efforts to utilize waste food for biogas have spread across Sweden. Signs in the Swedish Parliament's cafeteria advertise food scraps' second life as biogas.

The Linköping plant is unusual for its omnivorous appetite, also accepting human and packing-plant waste. This swill produces biofuel for buses, taxis, garbage trucks, and private cars, as well as a methane-propelled "biogas train" that runs between Linköping and Västervik on the southeast coast. The train's boosters (not squeamish vegetarians, from the sound of it) have figured that the entrails from one dead cow, previously wasted, buys 4 kilometers (2.5 miles) on the train.

The Danish Crown slaughterhouse uses the fat of 50,000 pigs in an average week to generate biogas. The entire Danish Crown plant has been redesigned with an eye to saving energy, part of a thirty-year Danish effort to eliminate waste, conserve energy, and reduce consumption of fossil fuels. Most of Denmark's energy infrastructure is owned by nonprofit cooperatives with resident shareholders. A majority of Denmark's people value high-quality health care, schools, and pensions over corporate profits, free individual choice, and low taxes.

## POND SCUM TO THE RESCUE?

In theory, the best source of biofuel may turn out to be algae—plain, ordinary pond scum. It has the additional cachet of consuming pollution as it supplies fuel. The theory is so enticing that in 2007 major-league money was being spent pursuing the idea by some electric utilities, notably Arizona Public Service, where an algae farm has been installed at its Redhawk power plant near Phoenix.

To hear its advocates, algae sounds like a near-perfect fuel source: It can be grown in almost any water, including brackish seawater and waste effluent. It needs only sunlight and carbon dioxide. Some algae also consume other pollutants. Others create starches that can be manufactured into ethanol. While an acre of corn can produce about 300 gallons of ethanol and an acre of soybeans can be turned into 60 gallons of biomass fuel, a single acre of algae, intensively cultivated, will produce a mind-numbing 5,000 gallons of biofuel a year, with no seasonal "down time" (Bourne 2007, 54, 57). A gallon of fuel produced from algae produces one-tenth the greenhouse gases of gasoline. Quite a bit of research and development lies between theory and reality for the pond-scum solution, however.

## REFERENCES

Andrews, Edmund L. 2007a. "Bush Makes a Pitch for Amber Waves of Home-grown Fuel." *New York Times*, February 23. http://www.nytimes.com/2007/02/23/washington/23bush.html.

Andrews, Edmund, and Felicity Barringer. 2007. "Bush Seeks Vast, Mandatory Increase in Alternative Fuels and Greater Vehicle Efficiency." *New York Times*, January 24. http://www.nytimes.com/2007/01/24/washington/24energy.html.

Associated Press. 2007a. "Biomass Fuel Plant in Iowa Is Moving Forward," *Omaha World-Herald*, October 5, D-1.

———. 2007d. "Germans Blame Ethanol Boom for—Oh Mein Gott!—Rising Beer Prices." *Omaha World-Herald*, June 3, A-18.

Ball, Jeffrey. 2007c. "Green-Fuel Alternative." *Wall Street Journal*, April 16, A-8.

Barrett, Joe. 2007. "Ethanol Reaps a Backlash in Small Midwestern Towns." *Wall Street Journal*, March 23, A-1, A-8.

Barta, Patrick. 2007a. "Crop Prices Soar, Pushing Up Cost of Food Globally." *Wall Street Journal*, April 9, A-1, A-9.

———. 2007b. "Jatropha Plant Gains Steam in Global Race for Biofuels." *Wall Street Journal*, August 24, A-1, A-12.

Bourne, Joel K. 2007. "Growing Fuel: The Wrong Way, the Right Way." *National Geographic* (October): 38-59.

Easterbrook, Gregg. 2007. "Global Warming: Who Loses—and Who Wins?" *Atlantic*, April, 5–64.

Environment News Service. 2007f. "Bush Orders First Federal Regulation of Greenhouse Gases." May 14. http://www.ens-newswire.com/ens/may2007/2007-05-14-06.asp.

———. 2007m. "Ethanol Production Threatens Plains States with Water Scarcity." September 21. http://www.ens-newswire.com/ens/sep2007/2007-09-21-091.asp.

———. 2007u. "Indy 500 Race Cars to Run on 100% Ethanol." May 25. http://www.ens-newswire.com/ens/may2007/2007-05-25-09.asp#anchor6.

———. 2007v. "Kansas Gets First U.S. Cellulosic Ethanol Plant." August 28. http://www.ens-newswire.com/ens/aug2007/2007-08-28-097.asp.

Etter, Lauren. 2007b. "With Corn Prices Rising, Pigs Switch to Fatty Snacks." *Wall Street Journal*, May 21, A-1, A-14.

Fahrenthold, David A. 2007. "'Green' Fuel May Damage the Bay: Ethanol Study Has Dire Prediction for the Chesapeake." *Washington Post*, July 17, B-1. http://www.washingtonpost.com/wp-dyn/content/article/2007/07/16/AR2007071601845_pf.html.

Fairless, Daemon. 2007. "Biofuel: The Little Shrub That Could—Maybe." *Nature* 449 (October 11): 652–55.

Fargione, Joseph, Jason Hill, David Tilman, Stephen Polasky, and Peter Hawthorne. 2008. "Land Clearing and the Biofuel Carbon Debt." *Science* 319 (February 29): 1235–38.

Goodell, Jeff. 2007. "The Ethanol Scam." *Rolling Stone*, August 9, 48–53.

Hansen, James E. Personal communication, April 12, 2007.

Hord, Bill. 2007a. "Closed-loop Ethanol Plant Plugged." *Omaha World-Herald*, December 1, D-1, D-2.

———. 2007b. "Mead Plant Hailed as 'Revolutionary.'" *Omaha World-Herald*, June 29, D-1.

Kennedy, Donald. 2007. "The Biofuels Conundrum." *Science* 316 (April 27): 515.

Krauss, Clifford. 2007. "Ethanol's Boom Stalling as Glut Depresses Price." *New York Times*, September 30. http://www.nytimes.com/2007/09/30/business/30ethanol.html.

Krugman, Paul. 2007. "The Sum of All Ears." *New York Times*, January 29, A-23.

Luhnow, David, and Geraldo Samor. 2006. "As Brazil Fills Up on Ethanol, It Weans Off Energy Imports." *Wall Street Journal*, January 9, A-1, A-8.

Monbiot, George. 2006a. *Heat: How to Stop the Planet from Burning.* Toronto: Doubleday Canada.

Mufson, Steven. 2006. "A Sunnier Forecast for Solar Energy; Still Small, Industry Adds Capacity and Jobs to Compete with Utilities." *Washington Post*, November 20, D-1. http://www.washingtonpost.com/wp-dyn/content/article/2006/11/19/AR2006111900688_pf.html.

———. 2007c. "In Microbe, Vast Power for Biofuel Organism's Ability to Turn Plant Fibers to Ethanol Captures Investors' Attention." *Washington Post*, October 18, D-1. http://www.washingtonpost.com/wp-dyn/content/article/2007/10/17/AR2007101702216_pf.html.

———. 2007e. "On Capitol Hill, a Warmer Climate for Biofuels." *Washington Post*, June 15, D-1. http://www.washingtonpost.com/wp-dyn/content/article/2007/06/14/AR2007061402089_pf.html.

Mufson, Steven, and Dan Morgan. 2007. "Switching to Biofuels Could Cost Lots of Green." *Washington Post*, June 8, D-1. http://www.washingtonpost.com/wp-dyn/content/article/2007/06/07/AR2007060702176_pf.html.

*New York Times.* 2007. "The High Costs of Ethanol." Editorial. September 19. http://www.nytimes.com/2007/09/19/opinion/19wed1.html.

Odling-Smee, Lucy. 2007. "Biofuels Bandwagon Hits a Rut." *Nature* 446 (March 29): 483.

*Omaha World-Herald.* 2007. "No Sugar-Beet Answer." Editorial. April 7, 6-B.

Patzek, Tad W. 2006. Letter to the editor. *Science* 312 (June 23): 1747.

Patzek, Tad W., and David Pimentel. 2005. "Thermodynamics of Energy Production from Biomass." *Critical Reviews in Plant Sciences* 24: 327–64. http://petroleum.berkeley.edu/papers/biofuels/uc_scientist_says_ethanol_uses_m.htm.

Richards, Bill. 2008. "A Good Combination: Biofuels, Smart Tilling." *Omaha World-Herald*, March 11, 7-B.

Rohter, Larry. 2006. "With Big Boost from Sugar Cane, Brazil Is Satisfying Its Fuel Needs." *New York Times*, April 10. http://www.nytimes.com/2006/04/10/world/americas/10brazil.html.

Rosenthal, Elisabeth. 2008. "Studies Deem Biofuels a Greenhouse Threat." *New York Times*, February 8. http://www.nytimes.com/2008/02/08/science/earth/08wbiofuels.html.

Searchinger, Timothy, Ralph Heimlich, R. A. Houghton, Fengxia Dong, Amani Elobeid, Jacinto Fabiosa, Simla Tokgoz, Dermot Hayes, and Tun-Hsiang Yu. 2008. "Use of U.S. Croplands for Biofuels Increases Greenhouse Gases Through Emissions from Land-Use Change." *Science* 319 (February 29): 1238–40.

Stone, Richard. 2007. "Can Palm Oil Plantations Come Clean?" *Science* 317 (September 14): 1491.

Strassel, Kimberly. 2007. "Ethanol's Bitter Taste." *Wall Street Journal*, May 18, A-16.

Thomas, Chris D. 2007. Review of Michael Novacek, *Terra: Our 100 Million Year Old Ecosystem. Nature* 450 (November 15): 349.

Vella, Matt. 2007. "Biofuel or Bust: On the Road with E-85." *Wall Street Journal*, June 19, D-2.

# CHAPTER 5

# The Power Forecast Is Windy

$B$y 2007, wind power was well past the realm of tree-hugger wishful thinking. It was cost-competitive with most fossil-fuel energy sources under some conditions, the fastest-growing source of energy in the United States and the world. By that year, wind power had become so popular that a shortage of parts was causing installations to fall behind demand. A new phrase in the English language, "wind rich," described an area with a relatively steady, unimpeded access to turbine-spinning breezes. Some surprising pitchmen for wind power (one of them being President George W. Bush) believe that the United States may derive as much as 20 percent of its electrical power from it by roughly 2020.

Wind power has room to grow in the United States. With electricity generation (49.7 percent coal-fired, 19.3 percent nuclear, 10.1 percent natural gas, 6.1 percent hydropower, 3.0 percent oil-fired) accounting for about 40 percent of U.S. greenhouse-gas emissions in 2007, wind power contributed 0.5 percent of electricity generation. Wind power in the United States in 2007 grew by 45 percent, adding 5,244 megawatts of capacity, a third of all the new electical capacity in the country. In that one year, wind energy employment in the United States doubled to about 20,000 (Smith 2008, A-6).

Wind power has become more scale adaptable, from huge turbines with blades the length of football fields to experimental household-scale units that have been widely criticized as overhyped and inefficient (and likely to shake a brick chimney to pieces). Turbines may be mounted on land or offshore; engineering is being developed to allow wind farms on water 200 meters (660 feet) deep.

Figures from more than seventy nations compiled by the Global Wind Energy Council revealed installation of 15,197 megawatts of wind power

**Figure 5.1. 750 kW NEG Micon Turbine in Moorhead, Minnesota.**
Source: National Renewable Energy Laboratory, Photographic Information Exchange. U.S.
Department of Energy.

during 2006, an increase of 32 percent compared to 2005, a year during
which the market grew by 41 percent. Total global wind-energy capacity
grew to 74,223 megawatts in 2006, up from 59,091 megawatts in 2005,
and 17,800 in 2000 (Environment News Service 2007q; Johnson 2007,
A-13). One megawatt provides power to about three hundred households.

Wind turbines have been popping up in some unexpected places. Parts
of the "Rust Belt" along the south shores of the Great Lakes are retooling
as wind-energy centers. Witness Lackawanna, New York, a suburb of Buf-
falo, where, according to a report in the *New York Times*, one sees

> on the 2.2-mile shoreline above a labyrinth of pipes, blackened buildings
> and crumbling coke ovens that was once home to a behemoth Bethlehem
> Steel plant: eight gleaming white windmills with 153-foot blades slowly
> turning in the wind near Lake Erie on a former Superfund site filled with
> iron and steel slag. (Staba 2007)

The wind farm is part of the New York State "brownfields" program, which
turns former low-level toxic sites to productive uses. To local boosters, the
Rust Belt has become the "Wind Belt." The turbines (called "Steel Winds")
cost $4.5 million each to build. Power lines once used by the steel plant now
carry electricity from the turbines, while paved roads, rail lines, and an industrial port built by Bethlehem are now wind-turbine infrastructure.

A wind turbine out back has become a corporate status symbol, a brandishing of "green" credentials. Wind turbines have been popping up all over the United States, many of them ad hoc corporate efforts. *Outside* magazine (a subsidiary of *Rolling Stone*) switched 90 percent of its Santa Fe, New Mexico, office's electricity to wind power. Burgerville, a fast-food chain in Washington and Oregon, by 2007 was using wind as supplemental power in its restaurants.

The New Belgium Brewery of Fort Collins, Colorado, during 1998, after a vote by employees, became the United States' first wind-powered beer factory. Employee-owners helped finance the change from their bonus pool. Recycling at New Belgium takes many forms, often turning "waste" products into products in creative ways. Spent grain, for example, is used as cattle feed. Used beer keg caps are used as table surfaces. The company also buys recycled materials whenever it can, from paper to office furniture. It saves electricity by using motion sensors on lights throughout the building, as well as induction fans that pull in cool winter air to chill its beer. New Belgium Brewery also delivers its beer in trucks burning biodiesel fuel (New Belgium n.d.).

## BIGGER AND BIGGEST

Various wind farms have competed for the title of biggest in the world, one leapfrogging another month by month. What was then being called the world's largest single wind-power generation facility was dedicated on September 26, 1998, near Lake Benton, Minnesota. Constructed, owned, and operated by the Enron Wind Corporation (a company that later went out of business), electricity generated by this wind facility was sufficient to supply power to 43,000 homes, replacing greenhouse gases equivalent to removing 50,000 cars from the road. The wind power was purchased by the Northern States Power Company for a service territory that includes much of Minnesota, as well as parts of Wisconsin, North and South Dakota, and Michigan.

By 2007, the Minnesota wind farm was a mere pygmy. In England, two enormous offshore wind farms were built in the outer Thames Estuary. The larger of the two, known as the London Array, claimed bragging rights as the largest offshore wind farm in the world. Soon thereafter, Mid-American Energy Company committed to spend $1.75 billion installing wind power across Iowa, enough to supply 300,000 homes, preventing emission of enough carbon dioxide to offset 43 percent of the state's motor vehicles. By 2007, Iowa already had 936 megawatts of installed wind capacity, third in the United States behind Texas (2,768) and California (2,361), according to the American Wind Energy Association (AWEA) (Gaarder 2007b, D-1). Minnesota had 895 megawatts and Washington 818.

Mid-American, a subsidiary of Warren Buffett's Berkshire Hathaway investment firm, profits from the sale of surplus energy and has not billed its customers for a rate increase since 1995. The company, which expects

**Figure 5.2. Cluster of Electrical Power Generating Wind Turbine on Rolling Hills, Altamont Pass, California.**
Courtesy of Shutterstock.

to hold rates stable until at least 2013, collected $735 million in federal tax credits for the wind installations, or about 40 percent of its project cost.

Some third-world countries are beginning to use wind power to reduce a small measure of their reliance on fossil fuels. India by the mid-1990s had installed 900 megawatts of wind-powered electrical energy and had plans to develop 500 megawatts of solar (photovoltaic) energy. India also concluded an agreement for a joint venture with Enron and Amoco to build a 50-megawatt solar-energy plant that will provide electricity to about 200,000 homes.

## WIND POWER GROWS AROUND THE WORLD

While wind power supplied a tiny fraction of total energy generated in the United States, some areas of Europe (Denmark, as well as parts of Germany and Spain) were using it as a major source. By 2003, Germany's northernmost state of Schleswig-Holstein was using wind for 28 percent of its electricity (Brown 2006, 201). Germany was the world leader in 2006 with 20,621 megawatts, 4.2 percent of electricity generation. Spain and the United States are in second and third place, each with a little more than 11,600 megawatts. India had 6,270 megawatts, and Denmark ranked fifth with 3,136. The United States led the world in new installations with 2,454 new megawatts installed during 2006 (Environment News Service 2007q).

Advances in wind-turbine technology adapted from the aerospace industry have reduced the cost of wind power from 38 cents per kilowatt-hour (during the early 1980s) to 4 to 8 cents, depending on wind conditions. This rate can be competitive with costs of power generation from fossil fuels, but costs vary according to site. Major corporations, including Shell International and British Petroleum, have been moving into wind power. Spain's tiny industrial state of Navarre, which generated no wind power in 1996, by 2002 generated 25 percent of its electricity that way. By 2007, 60 percent of its electricity was from renewable sources (mainly wind, with some solar), and it plans to raise that proportion to 75 percent by 2010.

In Schleswig-Holstein, the wind-energy industry has become the second largest employer after tourism. More than 30,000 wind-power-related jobs had been created across Germany by 2001, as private firms built rotors, towers, transfer stations, and ever-more-powerful turbines across wind-rich coastal states (Williams 2001, A-1). Denmark, a world leader in turbine technology, has turned wind power into a major earner of foreign exchange.

When Austin Energy, the publicly owned utility in Austin, Texas, launched its GreenChoice program in 2000, customers opting for green electricity paid a premium. During the fall of 2005, climbing natural gas prices pulled conventional electricity costs above those of wind-generated electricity, the source of most green power. This crossing of the cost lines is a milestone in the U.S. shift to a renewable energy economy.

"Strong growth figures in the U.S. prove that wind is now a mainstream option for new power generation," said Randy Swisher, president of AWEA. "Wind's exponential growth reflects the nation's increasing demand for clean, safe, domestic energy, and continues to attract both private and public sources of capital," he said. "New generating capacity worth US$4 billion was installed in 2006, billing wind as one of the largest sources of new power generation in the country—second only to natural gas—for the second year in a row" (Environment News Service 2007q).

Wind capacity in the Pacific Northwest, where it is often used in combination with hydroelectric power, has soared from only 25 megawatts in 1998 to a projected 3,800 megawatts by 2009. During 2006, Washington added 428 megawatts of wind power, trailing only Texas in new installations. According to Swisher, the electrical grid in the Northwest is especially inviting to wind-power developers because of its hydroelectric distribution network, relatively steady breezes, progressive utility companies, and new state laws that establish preferences for renewable energy.

Washington State law now requires utilities to work toward generating 15 to 25 percent of their electricity from sources that do not generate carbon dioxide. Utilities in 2007 planned to add 6,000 megawatts of wind power within a few years, more than doubling capacity. Hydroelectric dams built in river gorges are natural conduits for wind. The transmission lines connecting the dams with urban areas have been built with considerable surplus capacity—a built-in network for wind power. Farmers in the area have been earning $2,000 to $4,000 per year by allowing sites for wind turbines (Harden 2007, A-3).

Rapid development of wind power is worldwide. During 2006, Asia experienced the largest increase in installed capacity outside of Europe, with an addition of 3,679 megawatts, taking the continent over 10,600 megawatts, about half that of Germany. In 2006, the continent grew by 53 percent and accounted for 24 percent of new installations. China more than doubled its total installed capacity by installing 1,347 megawatts of wind energy in 2006, a 70 percent increase over 2005 (Environment News Service 2007q).

The Chinese market was boosted by the country's Renewable Energy law, enacted January 1, 2006. "Thanks to the Renewable Energy law, the Chinese market has grown substantially in 2006, and this growth is expected to continue and speed up," said Li Junfeng of the Chinese Renewable Energy Industry Association. "According to the list of approved projects and those under construction, more than 1,500 MW will be installed in 2007," Li observed. "The goal for wind power in China by the end of 2010 is 5,000 MW, which according to our estimations will already be reached well ahead of time" (Environment News Service 2007q).

## DENMARK'S EXAMPLE

Denmark was dependent on imported oil during the 1970s and made an enduring commitment to achieve energy independence when supplies were embargoed and its economy devastated. Wind power is an important part of this strategy, and as a result, Denmark has become a world leader in wind-turbine technology. Work on Danish turbines is a major reason why the technology today generates electricity that competes in price with oil, coal, and nuclear power. In the meantime, Denmark has built infrastructure that provides several thousand jobs.

In matters of advanced technology, Denmark dominates the worldwide wind-power industry; Danish companies have supplied more than half the wind turbines now in use globally, making wind-energy technology one of the country's largest exports. Some Danish wind turbines now have blades almost 300 feet wide—the length of a football field. During January 2007, a very stormy month, Denmark harvested 36 percent of its electricity from wind, almost double the usual.

In 2007, a center-right coalition that took power in Denmark reduced subsidies for wind-power development, and its growth slowed. Suddenly, the tax environment for new wind-power development was better in Texas (see below) than in Denmark. The Danes, however, had a very long head start.

Danish wind energy also has experienced technical setbacks, as wind operators taking advantage of stronger, steadier air currents, have built giant turbines at sea, some of which are more than 300 feet high, with blades nearly that wide. In one case, during 2004, turbines at Horns Reef, some ten miles off the Danish coast, broke down, equipment damaged by storms and salt water. Vestas, a Danish manufacturer, fixed the problem by replacing the equipment at a cost of $38 million. But Peter Kruse, the head of investor relations for Vestas, warned that the lesson from Horns Reef was that wind farms at sea would remain far more expensive than those on land. "Offshore wind farms don't destroy your landscape," Kruse said, but the added installation and maintenance costs are "going to be very disappointing for many politicians across the world" (Kanter 2007).

Technical and natural difficulties aside, wind power has become a major source of electricity in Denmark. Some western parts of the country derive 100 percent of their peak needs from wind if a stiff breeze coincides with demand. If higher governmental subsidies had been maintained, Denmark could now be generating close to one-third—rather than one-fifth—of its electricity from windmills (Kanter 2007). The Danish wind-power industry also provides 150,000 families with shares of profits from the national electricity grid. Thousands of rural Danish residents have joined wind-power cooperatives, buying turbines and leasing sites to build them, often on members' land (Woodard 2001, 7). At the same time, power companies in Denmark are now taxed for each ton of carbon dioxide emitted above a low (and, over time, gradually declining) limit.

Colin Woodard, writing in the *Christian Science Monitor*, described a Danish wind farm:

> Arriving in the Danish capital by sea, the visitor's first glimpse of the country is the Middelgrunden Wind Farm, a row of 20 enormous wind turbines gently spinning above the waves nearly two miles offshore. Completed in December [2001], Middelgrunden is the world's largest offshore wind farm, dominating the views from Copenhagen's docks and seaside parks. With blades 100 feet long, its wind machines generate enough power to supply 32,000 households. It's a fitting introduction to Denmark, a nation of 5 million that has emerged as the undisputed leader in wind energy. (2001, 7)

According to another on-site observer, "A graceful arc of alabaster windmills rises out of the midnight-blue waters of the Oresund strait, visual testimony to Danes' commitment to clean energy and the health of the planet" (Williams 2001, A-1).

While Denmark has made impressive progress, the country is not as green as its wind-power figures indicate. Most of Denmark's other power comes from generating plants that burn imported coal. "We are losing ground," said Anne Grete Holmsgaard, the energy spokeswoman for the opposition Socialist People's Party in Denmark. "It's terrible, actually, that we're not that green as we should be" (Kanter 2007).

## VISUAL POLLUTION, BIRD-KILL, AND WIND FICKLENESS

Wind turbines have not been universally welcomed. Some of their neighbors complain that the larger wind farms comprise "visual pollution." The book *Cape Wind* (Williams and Whitcomb 2007) described a battle over siting of wind turbines on Massachusetts's Cape Cod, specifically the Cape Wind farm on Nantucket Sound. Perhaps a coal strip mine would be a less salubrious sight, but no one receiving coal-fired energy on Cape Cod has to witness the mining of the coal. The Cape Wind project is no garden-variety project. Its plans include 130 turbines producing 3.6 megawatts each. Each machine uses a 175-foot blade. On a windy day, at maximum capacity, all 130 machines could produce as much power as a modest-size electricity plant burning coal or natural gas.

The power capacity of wind turbines may be restricted only by the size of a blade that can be hauled to a site, mounted, and maintained with some degree of structural integrity. A turbine at 250 feet takes advantage of winds that are 20 percent stronger at that elevation than at 150 feet, according to. Mark Z. Jacobson, an associate professor at Stanford University's Department of Civil and Environmental Engineering (Wald 2007). The bigger the turbine, though, the greater the chance it may kill birds—about 40,000 of them a year in the United States by 2006. That is only one bird per thirty turbines per year, however, a small fraction of the millions killed each year by domestic cats (Marris and Fairless 2007, 126).

Another major problem for wind energy is a matter of timing. The hottest days (when power demand is highest, for air conditioning) are usually the least windy. "As a result, wind turns out to be a good way to save fuel, but not a good way to avoid building plants that burn coal," wrote Matthew Wald in the *New York Times*. He continued:

> A wind machine is a bit like a bicycle that a commuter keeps in the garage for sunny days. It saves gasoline, but the commuter has to own a car anyway. Without major advances in ways to store large quantities of electricity or big changes in the way regional power grids are organized, wind may run up against its practical limits sooner than expected. In many places, wind tends to blow best on winter nights, when demand is low. (Wald 2006)

The timing problem can be surmounted with storage systems that are available. Robert E. Gramlich, a policy director with AWEA, also suggests that wind energy can be integrated into an electrical grid on a large enough area that ebbing wind in one part will be balanced by continuing breezes in another.

## WIND POWER IN TEXAS OIL COUNTRY

While Texas has a reputation as a conservative state run by oil barons, it has become one of the most progressive environments for wind energy in the United States. George W. Bush, himself once an oilman, helped start the "wind rush" when he was governor by signing a bill requiring the state's power infrastructure to include 3 percent from alternative energy by 2009. In December 2004, state planners recommended that by 2025 10 percent of Texas's power should come from renewable sources.

Cielo Wind Power of Austin has been buying sixty-year "wind rights" from ranchers in Texas with plans to build the Noelke Hill Wind Ranch for $130 million. The wind ranch is expected to generate 240 megawatts of power, enough to provide electricity to 80,000 homes through TXU, a utility based in Dallas. Several companies, among them American Electric Power and General Electric, have plans to spend as much as $1 billion on wind energy in the midst of the Permian Basin, which has heretofore been known as oil country. Wind speed in the area averages 16 miles an hour.

By late 2002, Cielo had built $300 million worth of wind-power turbines in Texas and sold them to various power companies, "jutting more than 200 feet into the air with blades 200 feet in diameter, to capture high-velocity wind patterns" (Herrick 2002, B-3). Texas by late 2002 was generating enough wind power to supply about 300,000 homes (Herrick 2002, B-3). A standard line used by wind developers with the ranchers in the area is: "You've been getting your hat blown off your head your whole life. It's time to stop cussin' and make some money" (Herrick 2002, B-3). Wind turbines placed roughly one per 25 acres earn the average landowner about $3,200 each (Herrick 2002, B-3).

On October 1, 2006, Texas's Republican governor, Rick Perry, announced a $10 billion public-private initiative to expand wind energy. He proudly pointed out, with a trademark Texas penchant for bragging rights, that his state had surpassed California as the U.S. leader in wind-generation capacity. This most recent initiative was expected to add 10,000 megawatts of wind power within the state, enough to supply power to 2.4 million homes (Environment News Service 2006d). The program that Perry announced allows private companies to supply capital for wind farms, while the state's Public Utility Commission directs construction of transmission lines to capture and deliver power. "I am proud of our state's commitment to renewable energy production," Perry said. "We are on the leading edge of developing renewable sources of energy and a more diversified energy economy which is key to keeping costs down" (Environment News Service 2006d).

Perry pointed out that every thousand additional wind-generated megawatts will reduce carbon-dioxide emissions by six million tons and will create jobs in Texas—much the same rationale that built wind power into a world-class industry in Denmark. "With this $10 billion announcement, the economic ripple will be more like a tidal wave as these companies pour millions of dollars into wages and salaries for Texas workers," Perry declared. "This is a monumental investment that will make our air cleaner and our people healthier" (Environment News Service 2006d).

A year later, wind was becoming truly big business in Texas. By 2007, Royal Dutch Shell and a wind-development company owned by Goldman Sachs Group were building wind-power infrastructure in the Texas Hill Country around Silverton. The companies were building some of the largest wind farms in the world, planning hundreds of turbines costing about $2 million each (Ball 2007d). The area, once home to oil rigs, has a relentless wind that blows so steadily that the local flora (sagebrush and mesquite) grow at an angle. The area's canyons also tend to funnel the wind.

In the same area, Shell, a company diversifying from its oil roots, was planning a 120-square-mile wind farm, still another that was being touted as "the largest in the world," occupying a land area five times the size of Manhattan Island. Shell's new wind farm will produce electricity equal to that of a coal-fired power plant (Ball 2007d, A-11).

Shell and other companies are betting that wind power's time has come. Their lobbying contacts in Austin, the state capital, are poised to make a reality of this hunch. Large companies, such as General Electric, which manufactures wind turbines, also have joined the wind lobby. Presently, federal tax rules allow a company engaged in wind-power generation to reduce its tax bill by 1.9 cents per kilowatt-hour. Congress has authorized and reauthorized the credit several times (usually two years at a time). The short-term nature of this federal tax subsidy has inhibited the construction of wind infrastructure, however.

By 2007, about twenty states also allowed tax preference for wind energy. Texas law further requires utilities to buy wind power when and where it is available as a proportion of generating capacity. Shell by 2007 was selling wind power from its 30-square-mile Brazos wind farm to TXU, which is under pressure by environmentalists to reduce its reliance on coal-generated power. The Brazos wind farm contains 160 turbines that have a total capacity of 160 megawatts. Local ranchers can collect as much as $80,000 a year from energy companies for a 640-acre section of land under a wind farm (Ball 2007d, A-11).

## HOUSEHOLD WIND TURBINES

A growing number of people in the United States have been installing household-scale wind-power generators, which are available for $10,000 to $15,000. Herschel Carter, 62, a retired insurance agent, cut the amount of electricity he consumed from the grid nearly by half (1,752 to 993 kilowatt-hours) in the first month his wind turbine operated (March 2007). His electric bill for that month fell $68. Payback will take several years, but Carter regards the turbine as a form of energy security because his wife is on oxygen that requires electricity. Their household turbine has a battery that stores energy on windy days for use when wind is calm (McNulty 2007, 7).

Household-scale wind turbines were invented in the United States about 1920. Presently, the country remains a world leader in their technology; more than 90 percent of household wind turbines have been installed here. Growth in the number of household wind turbines has been averaging 14 to 25 percent a year between 2000 and 2006 worldwide, according to AWEA executive director Swisher. At least a half-acre of land is required for a residential wind turbine, and regulators must permit a tower of at least 35 feet (the higher the tower, the more efficient the turbine).

Smaller wind turbines may be available. A miniature wind turbine developed by the Scottish company Windsave can be mounted on an urban roof at a cost of about $2,000. Its four-foot-diameter blades can provide part of an average home's electricity (company publicity optimistically asserts one-third) (McGuire 2005, 175).

**Figure 5.3. A household wind turbine.** © Olivier Le Queinec.
Courtesy of Shutterstock.

Small-scale wind power advocates have been severely criticized for over-stating potential savings. Windsave, in particular, has been savaged by crit-ics for stating on its website that its machine, mounted on a roof, can save an average home 25 to 40 percent of residential energy costs. Critics find that the strength of the wind and the power of the turbine both have been exaggerated for sales purposes. And the calculations are for a British home, which uses about half the electricity of an average home in the United States. Similar sales tactics have been criticized in reference to U.S. manu-facturers such as Skystream. A wind turbine with a rotor that is six feet or less in diameter will not save a significant portion of most Americans' energy consumption, even in a windy area. It's simply too small to pro-duce enough energy with present technology. Roofs also tend to obstruct wind, defeating the turbine's purpose unless it is mounted several feet above a roof's peak.

Critics of household wind energy raise other concerns. For example, many U.S. small turbines have no mechanism to stop a machine in an overspeed, spinning too quickly (usually as a result of very strong storm winds) until it destroys itself. Danish wind turbines have been required to build in stopping mechanisms for decades. Any wind turbine also transmits vibration to any structures on which it is mounted. In a few instances, small wind turbines have worked their way off roofs during storms and plunged into homes, causing damage. Others have reduced brick chimneys to rubble.

# REFERENCES

Ball, Jeffrey. 2007d. "The Texas Wind Powers a Big Energy Gamble." *Wall Street Journal*, March 12, A-1, A-11.

Brown, Lester R. 2006. *Plan B: Rescuing a Planet under Stress and a Civilization in Trouble*, rev. ed. New York: Earth Policy Institute/W. W. Norton.

Environment News Service. 2006d. "Texas Announces $10 Billion Wind Energy Deal." October 3.

———. 2007q. "Global Wind Power Generated Record Year in 2006." February 12.

Gaarder, Nancy. 2007b. "Winds of Change Blowing in Iowa." *Omaha World-Herald*, May 13, D-1, D-2.

Harden, Blaine. 2007. "Air, Water Powerful Partners in Northwest; Region's Hydro-Heavy Electric Grid Makes for Wind-Energy Synergy." *Washington Post*, March 21, A-3. http://www.washingtonpost.com/wp-dyn/content/article/2007/03/20/AR2007032001634_pf.html.

Herrick, Thaddeus. 2002. "The New Texas Wind Rush: Oil Patch Turns to Turbines, as Ranchers Sell Wind Rights; A New Type of Prospector." *Wall Street Journal*, September 23, B-1, B-3.

Johnson, Keith. 2007. "Alternative Energy Hit by a Windmill Shortage." *Wall Street Journal*, July 9, A-1, A-13.

Kanter, James. 2007. "Across the Atlantic, Slowing Breezes." *New York Times*, March 7. http://www.nytimes.com/2007/03/07/business/businessspecial2/07europe.html.

Marris, Emma, and Daemon Fairless. 2007. "Wind Farms' Deadly Reputation Hard to Shift." *Nature* 447 (May 10): 126.

McGuire, Bill. 2005. *Surviving Armageddon: Solutions for a Threatened Planet*. New York: Oxford University Press.

McNulty, Sheila. 2007. "U.S. Power Generation Answer Is Blowing in the Wind." *Financial Times* (London), April 24, 7.

New Belgium. N.d. "Our Story." http://www.newbelgium.com/story.php.

Smith, Rebecca. 2008. "Wind, Solar Power Gain Users." *Wall Street Journal*, January 18, A-6.

Staba, David. 2007. "An Old Steel Mill Retools to Produce Clean Energy." *New York Times*, May 22. http://www.nytimes.com/2007/05/22/nyregion/22wind.html.

Wald, Matthew. 2006. "It's Free, Plentiful and Fickle." *New York Times*, December 28. http://www.nytimes.com/2006/12/28/business/28wind.html.

———. 2007. "What's So Bad about Big?" *New York Times*, March 7. http://www.nytimes.com/2007/03/07/business/businessspecial2/07big.html.

Williams, Carol J. 2001. "Danes See a Breezy Solution: Denmark Has Become a Leader in Turning Offshore Windmills into Clean, Profitable Sources of Energy as Europe Races to Meet Emissions Goals." *Los Angeles Times*, June 25, A-1.

Williams, Wendy, and Robert Whitcomb. 2007. *Cape Wind: Money, Celebrity, Class, Politics, and the Battle for Our Energy Future on Nantucket Sound*. New York: PublicAffairs.

Woodard, Colin. 2001. "Wind Turbines Sprout from Europe to U.S." *Christian Science Monitor*, March 14, 7.

# CHAPTER 6

# Harvesting the Sun

Solar power has advanced significantly since the days of inefficient photovoltaics. In California, solar power is being built into roof tiles, and talk
is that nanotechnology will someday allow any surface that the sun hits to
become a source of power—windows, for example. Experiments have been
undertaken with a new form of solar energy—Concentrating Solar Power
(CSP)—that uses mirrors to harvest the sun on a scale that will match the
scale of power plants fed by fossil fuels. In our lifetimes, as private homes
using alternative energy sources feed power into the electrical grid, meters
may run backward, paying householders for contributed power. The day
may come when carbon-based fuels are used only as backup for periods of
peak demand when a medley of other sources do not satisfy needs.

"Solar power has captured the public imagination," wrote Andrew C.
Revkin and Matthew L. Wald in the *New York Times*.

> Panels that convert sunlight to electricity are winning supporters around the
> world—from Europe, where gleaming arrays cloak skyscrapers and farmers'
> fields, to Wall Street, where stock offerings for solar panel makers have had a
> great ride, to California, where Gov. Arnold Schwarzenegger's "Million So-
> lar Roofs" initiative is promoted as building a homegrown industry and
> fighting global warming. (Revkin and Wald 2007)

For all the excitement, however, solar power in 2006 contributed only
0.01 percent of the United States' electricity supply. At the same time,
worldwide, one coal-fired electric plant was being built *every week*. "Most
of the environmental stuff out there now is toys compared to the scale we
need to really solve the planet's problems," said Vinod Khosla, a prominent Silicon Valley entrepreneur who focuses on energy (Revkin and Wald
2007).

Solar is becoming increasingly efficient for home-scale electric power. For several years, university students and sponsors from the United States, Canada, Germany, and Spain have answered a challenge from the National Renewable Energy Laboratory of the U.S. Department of Energy (DOE) to build comfortable and affordable 800-square-foot prototype solar houses on the National Mall in Washington, D.C. About 100,000 people visit the ten-day Solar Decathlon. Many of the houses are so efficient that they have solar-generated electricity left over to power electric cars. Germany's Technische Universität Darmstadt took top honors in the 2007 Solar Decathlon competition.

Until recently, photovoltaic cells were able to convert only about 20 percent of the solar energy they received into energy. By 2007, however, a team of researchers at the University of Delaware had converted 42.8 percent in an experimental project (a prototype had yet to be built at that time). They were aiming at 50 percent. The work was being done under a $12 million Department of Defense contract to develop battlefield electronics. The group uses a novel light-splitting technique that increases efficiency markedly. While the concept is viable in the laboratory, cost has been a major problem. During 2007, however, the DuPont company put $100 million on the line to make the concept commercially viable (Kintisch 2007a, 583–84).

The solar-power industry provided about 20,000 jobs across the United States in 2006, according to Rhone Resch, president of the Solar Energy Industries Association. That's a small number in a country of 300 million people, but Resch said solar-sector jobs are growing by 35 percent a year. Many of the jobs also are good-paying. "You're producing high-quality manufacturing jobs when others are moving out of the United States," said Resch. "If you look at the next high-tech growth industry in the United States, it can and should be solar energy" (Mufson 2006, D-1). According to the DOE, worldwide shipments of photovoltaic cells more than doubled, from slightly less than 800 megawatts to more than 1,700 megawatts per year, between 2003 and 2005.

Some new housing developments already have been designed to stop the electric meter completely, with enough solar power from their roofs to supply a home's every necessary kilowatt. Premier Gardens in the Sacramento, California, area touted "zero-energy homes" in a subdivision that produced 300 megawatts of electricity a year, using photovoltaic roof tiles with integrated solar technology developed and sold by General Electric (Rogers 2006, 2-RE).

## SOLAR POWER WORLDWIDE

Technology is improving steadily to make affordable what was once a showpiece source of energy. Sometimes solar power leaps the fossil fuel age. Consider Tecnosol, a small company in Nicaragua, where more than half of rural people have no access to electricity. In some areas, most notably in the remote eastern provinces, firewood is still the main source of fuel for cooking. Many people suffer respiratory diseases from wood smoke or spend scarce money on kerosene. Tecnosol in 2007 was installing 25,000 solar units, which will cut carbon dioxide emissions by 150,000 tons over the life of the equipment (Mallaby 2007, A-17).

As with wind, the builders of solar plants these days have been claiming superlatives, leapfrogging each other with bragging rights to the biggest and best. By the end of 2007, for example, Nellis Air Force Base in Nevada was poised to open what its publicists called the largest solar-power complex in the United States.

Los Angeles–based Solar Integrated Technologies struck a deal during January 2007 with the British supermarket chain Tesco to build the world's largest (at that time) rooftop solar panel installation. Solar Integrated said that it had won a $13 million contract to install panels atop Tesco USA's new distribution center in Riverside, California. The system will be designed to provide a fifth of the depot's power supply.

Denmark hosts a 12-megawatt district heating station that runs on solar power on the island of Aero, also billed by its builders as the largest such station in the world. The plant's photovoltaic cells are so sensitive that they harvest sunlight even on Denmark's many cool, cloudy days.

In England, Susan Roaf's solar roof fuels her electric car. Roaf, an architect at England's Oxford Brookes University, has designed a solar house fitted with photovoltaic cells that harvest the sun's energy and convert it into electricity. According to a report in the London *Times*, "Her

**Figure 6.1. 15 megawatt solar photovoltaic array at Nellis Air Force Base, Nevada, completed December 2007.**
Source: U.S. Air Force [http://www.nellis.af.mil].

system is so efficient that she uses it to charge her electric car's batteries and makes a profit by selling 57 percent back to the national network" (O'Connell 2002). Using similar technology, cities could someday become self-sufficient power generators without use of fossil fuels. "Once [solar] fuel cells have been perfected, we could all own one," Roaf said. "If we were to convert our homes to solar power-using solar panels to heat the house, and [photovoltaic] cells to convert the sun's energy into electricity, we could not only store the excess for a rainy day but could sell it back to the National Grid. This would be no bad thing." The British government has made £20 million available in the form of grants for conversion to solar energy. Currently photovoltaic cells are very expensive—Roaf's roof cost £25,000 (about $50,000) in 1995—but the price will decrease as more people become energy self-sufficient. "After all, look at it this way," she said. "The roof is earning its keep" (O'Connell 2002).

Despite its often-stated reticence about global warming, the George W. Bush administration in January 2003 allowed the National Park Service to install the first-ever solar electric system on the grounds of the White House. The Park Service, which manages the White House, installed a nine-kilowatt rooftop solar electric photovoltaic system, as well as two solar thermal systems that are used to heat water.

In some northern Chinese farming villages, homes that a few years ago relied solely on low-grade coal for heat and cooking now utilize solar-heated rooftop water tanks (Landauer 2002, F-4).

About 85 percent of Israeli households have solar water heaters, which the government estimates lighten the country's overall energy burden by 3 percent. Solar heaters have been used in Israel since the 1950s. The global energy crisis of 1979 reminded Israelis of their reliance on foreign sources for oil and coal. Solar water heaters have been required in new homes by law in Israel since 1980. About twenty companies, employing four thousand people, manufacture the systems and sell them for $300 to $1,000 each. A midrange system can pay for itself in energy savings in three or four years and, if maintained, can last more than ten years (Kaplow 2001, 1-P).

Yet another larger-than-thou claim was posted in 2007 by the Fresno Yosemite International Airport, as it planned to install the largest solar electric project at any airport in the United States. Fresno's mayor Alan Autry said, "This project further establishes Fresno as a national leader amongst municipal governments in the innovative use of renewal energy and protecting the environment." WorldWater and Power Corporation, developer and marketer of proprietary high-power solar systems, has been awarded a twenty-year solar electric power purchase contract by the Fresno City Council for the airport solar system (Environment News Service 2007o). When complete in 2008, the airport's solar platform will cover 25 acres and generate two megawatts, saving the airport $13 million a year in energy costs.

Google early in 2007 began building a rooftop solar-powered electricity system at its Silicon Valley headquarters, which will be the largest U.S. solar-powered corporate office complex. The system generates

1.6 megawatts, enough, under other circumstances, to power about a thousand California homes. The announcement of the new installation was, once again, an occasion for superlatives. "This is the largest customer-owned solar electric system at a corporate site," said Noah Kaye, director of public affairs at the Solar Energy Industries Association, an industry group based in Washington, D.C. (Auchard and Anderson 2006).

Google will use solar power for almost a third of the electricity consumed by office workers at its headquarters, excluding power consumed by data centers that power many of Google's Web services worldwide. Data centers usually consume about ten times more electricity than buildings that house office workers. "We are going to be producing roughly 30 percent of the power that we use," David Radcliffe, vice president of real estate at Google, told Reuters (Auchard and Anderson 2006).

## CONCENTRATING SOLAR POWER (CSP)

A new type of solar technology, Concentrating Solar Power, is much more powerful than photovoltaic cells. A rooftop photovoltaic complex might power a small office building, while a CSP complex near Seville, in southern Spain, can generate 11 megawatts, enough electricity for about six thousand homes (Abboud 2006, A-4). A similar-sized development was under way by 2008 in the Mojave Desert of California, designed ultimately to serve about 400,000 homes. The CSP mirrors track the sun and concentrate its power on a single point, generating steam that runs a turbine. Some of the heat also is stored in oil or molten salt to run the turbine after sunset or when clouds obscure the sun. Such new technologies may increase the potential of solar power and bring down its cost, now 12 to 15 cents per kilowatt-hour, compared to an average of 4 cents for coal-fired energy.

According to the consulting firm Emerging Energy Research, forty-five CSP projects were in planning stages around the world by 2007. The Spanish government has set a goal of 500 megawatts of solar power by 2010. Spain is presently subsidizing CSP development, requiring utility companies to buy their power at above-market rates. The solar company Abengoa plans by 2013 to build enough CSP capacity to supply all of Seville, about 180,000 homes.

The 11-megawatt plant near Seville was underwritten by the regional government of Andalusia and Abengoa, whose parent company, Solúcar, built the power plant. The power plant in the municipality of Sanlúcar la Mayor, fifteen miles west of Seville, produces electricity with 624 large movable mirrors called *heliostats* (Environment News Service 2007n). The site includes a 380-foot concrete tower surrounded by the huge mirrors. Each CSP mirror's surface measures 120 square meters (1,290 square feet). They concentrate the sun's rays atop the tower with a solar receiver, and a steam turbine drives an electricity generator. This CSP plant is the first of several solar power generation stations planned for the same area, which will generate more than 300 megawatts by 2013, using both CSP and conventional photovoltaics (Environment News Service 2007n). The

**Figure 6.2. Concentrating Solar Power (CSP) Stirling Energy Systems, Inc. (SES)/Boeing, 25 kW Dish Stirling system at sunset.**
Source: National Renewable Energy Laboratory, Photographic Information Exchange. U.S. Department of Energy.

Sanlúcar la Mayor Solar Platform will replace emission of more than 600,000 metric tons of carbon dioxide per year.

The Spanish project is partially funded by the European Union, which has been supporting solar development for more than ten years. The European Commission also publishes maps of Europe's solar power potential. These maps, produced by the Photovoltaic Geographical Information System of the Joint Research Centre, also include an interactive service allowing users to calculate the solar power potential of any location in Europe (Environment News Service 2007n).

In the desert north of Tucson, Arizona Public Service has been experimenting with CSP, using an array of mirrors that focuses sunlight and heats mineral oil to 550°F; the heat then evaporates a liquid hydrocarbon, which runs a generator to make electricity. The array includes six rows of mirrors, each nearly a quarter-mile long, providing about 100,000 square feet of reflective space. This demonstration project produces one megawatt of power, enough to power a large shopping center. Acciona Solar Power, the company that installed the experimental array, in 2007 was planning a 350-acre plant near Boulder City, Nevada, with a capacity of 64 megawatts, also using

**Figure 6.3. A home-scale photovoltaic system.**
Source: National Renewable Energy Laboratory, Photographic Information Exchange. U.S. Department of Energy.

CSP. Arizona Public Service and several other utilities also were considering a joint project to build a 250-megawatt CSP plant (Wald 2007).

## PASSIVE SOLAR POWER

Solar-conscious homeowners do not need to wait for a CSP plant to arrive in their neighborhoods. Solar power comes in several varieties, some of them immediately available and quite prosaic. The most elemental is passive. Georg Zielke and his wife and children share a five-bedroom "passive house" in Darmstadt, Germany, with heating costs 90 percent lower than their neighbors'. Extra insulation and state-of-the-art ventilation recycle the energy from passive sources such as body heat, the sun, and household appliances to warm the air. When it gets really cold, the Zielkes just turn on the TV.

The German government has thrown its weight behind the idea, guaranteeing low-cost loans for people who want to utilize a passive solar energy system at home. Invented in a German-Swedish joint-venture in the early 1990s, about ten thousand have been built in Europe so far, most of them in Germany.

## COST OF SOLAR POWER GENERATION DECLINES

Until CSP becomes commonplace, the most widespread type of solar power will continue to utilize photovoltaic cells, which require no fuel (other than the sun) and little maintenance. The conversion of solar energy to electricity takes place with no moving parts and causes very little environmental disturbance. Because the sun must be shining to produce energy, however, it is likely that photovoltaic solar will develop, at least initially, with backup power from existing power-generation networks.

By the 1990s, photovoltaic cells already were less expensive as a source of electricity and hot water than extension of conventional, centralized power distribution to some remote locations, such as weather stations. Their efficiency has vastly improved, and that trend probably will continue through the twenty-first century. By the end of the century, some present-day observers expect that many urban homes will produce their own electricity through solar circuits built into rooftops. Centralized, fossil-fueled electricity generation may become only one of several choices for electric-power generation.

During the 1970s, photovoltaic cells were manufactured by sawing large crystals of silicon into thin slices, an expensive and inefficient process. By the 1990s, however, work was under way to produce photovoltaic energy from semiconductor (computer) chips, potentially at a much lower cost once the technology is refined.

## APPROPRIATE-SCALE TECHNOLOGY

Solar power creates opportunities for decentralized "appropriate-scale" technology, especially in countries with large rural populations. One example is India, which averages 210 days a year of nearly direct sunlight, a large rural population, and a tradition of local, basic, small-scale problem solving that stems from Mahatma Gandhi, who turned homespun cloth from a small spinning machine into a powerful political symbol vis-à-vis the centralized weaving industry controlled by the British.

By 1995, six thousand villages in India that had no access to conventional electric power grids were drawing electricity from banks of photovoltaic solar cells. The number of such arrays doubled during the next decade. Using the same model of small-scale, locally controlled technology, photovoltaic modules and solar cooking stoves are being used increasingly in India's villages. Many villagers also use biogas digesters that convert the dung of cows and other animals to energy. The resulting methane is burned as energy before it reaches the atmosphere as a greenhouse gas. New technology also allows dung to be turned to an energy-rich sludge without smoke and fire. A million digesters were operating in India by 1990, despite the fact that one of them costs about $50 (with half the amount paid by a government subsidy)—almost one-fifth of the average rural Indian's annual cash income (Oppenheimer and Boyle 1990, 137, 139). The digester-financing program is administered by the Indian federal government's Department of Nonconventional Energy.

These programs should not leave an impression that India, as a whole, is reducing its greenhouse-gas emissions, unfortunately. In India's cities, a growing middle class is creating demands for more energy, most of which is generated from fossil fuels, especially coal. India has only 1 percent of the world's coal reserves, but it is fourth among the world's nations in coal combustion. In twenty years (1985 to 2005), India's electrical generating capacity multiplied several times, and it still lags demand severely. Electricity generation by fossil fuels is still growing rapidly on the Indian subcontinent, even with recent solar and wind power development.

## SOLAR POWER DESIGNED INTO NEW HOMES AT A DISCOUNT

With technological improvements bringing the cost of solar energy down to between $7 and $10 per watt in 2006 (compared with about $20 in 1995), more new homes across the United States were being built with solar panels embedded in their roof tiles that sometimes can supply a substantial fraction of an average home's electricity needs. Johnson Square Village in Brockton, Massachusetts, for example, installed photovoltaic cells in its roof tiles and saved each condominium owner about $600 a year in electricity costs. In some areas, electricity meters were being installed that could run backwards, crediting consumers for energy accumulated when their solar systems produced more energy than their homes consumed. Federal and state tax breaks (California and New Jersey had the most generous) made solar energy more affordable.

As California architects were designing solar systems into custom homes, the *New York Times* reported early in 2007 that solar power was becoming something of a status symbol in California, "a glamorous mark of personal responsibility. Celebrities, including Leonardo DiCaprio, Alicia Silverstone, Carlos Santana, and Tom Seaver, have installed solar systems. Edward Norton runs a campaign in Los Angeles encouraging his fellow celebrities to install solar panels on their homes and to make donations for systems in low-income housing" (Dicum 2007).

Living in Scottsdale, Arizona, Bud Annan, solar program director at DOE during the Clinton administration, has been working with utility and local real-estate developers to try to incorporate solar roofs into 10,000 new houses, all at once. That way, he said, the installers can go from house to house the way carpenters, plumbers, and electricians do. "He can standardize his installation, and that whole second half of the equation becomes more manageable for him," Annan said. This is much more cost-effective than single installations (Wald 2007).

California's state legislature, having approved the California Solar Initiative, offered homeowners a rebate on top of a federal tax credit of up to $2,000 to install solar systems. Banks now often factor in a solar system as an improvement that increases a house's value. Companies such as NextEnergy provide homeowners with a package including system design, permit applications, rebate processing, installation, maintenance, and a warranty.

By 2007, a combination of state and federal tax credits could reduce the retail cost of a home solar-power system by 40 to 60 percent. In California, a $2.80-per-watt credit, plus a 7.5 percent tax credit and a $2,000 federal tax credit, pays about 40 percent of the cost. In Colorado, a state tax credit of $4.50 per watt, plus the $2,000 federal credit, pays about 60 percent. New York and New Jersey credit at a rate close to Colorado's. A solar system for a 2,500-square-foot home will cost $12,000 to $15,000 after credits, depending on the state (Steinman 2007, 280).

In many areas of the United States, power companies now are legally required to credit customers for surplus power that they produce. The grid, in effect, serves to store power, replacing the bank of batteries that is a component of off-grid systems. At the end of the year, credits for solar power added to the grid are applied against charges for power taken from it, helping homeowners "zero out" their electricity bills (Dicum 2007).

## SOLAR POWER IN CLOUDY GERMANY

Opened near Espenhain, Germany, during 2004 in an area that had been a dumping ground since the 1930s for millions of tons of coal-mine waste produced in nearby mines, a Geosol solar plant for a time claimed bragging rights as the largest of its kind in the world. It was said to be "so clean and green that it produces zero emissions and so easy to operate that it has only three regular workers: plant manager Hans-Joerg Koch and his two security guards, sheepdogs Pushkin and Adi" (Whitlock 2007). Even northern Germany, a land of fog and weeping skies, was finding a place for solar cells so sensitive that a hot shower on a cloudy day was not remarkable. By 2006, half the world's new solar capacity was being installed in Germany, despite the fact that it has only half as many sunny days as Portugal, a more obvious solar success story (Whitlock 2007). The German solar panels work on drizzly days, although they generate only a quarter to half the electricity as a sunny day produces.

Coal mining was Espenhain's largest employer, providing eight thousand jobs under the East German regime that collapsed during 1989. After German reunification, the mining jobs vanished.

> "This region was known as the dirtiest in all of Europe," said Juergen Frisch, mayor of Espenhain. "The solar plant came at a very good time for Espenhain. It's helped to change our image." Unlike the coal mines, the solar plant makes almost no noise, save for the low thrum of a few outdoor air-conditioning units that cool the electrical transformers. The plant, with 33,500 solar panels, sits on a 37-acre site in a field off a rural road and requires scarcely any maintenance. (Whitlock 2007)

A German law enacted in 2000 requires the country's utility companies to subsidize new solar installations by buying their electricity and using it on their grids at marked-up rates that allow small solar enterprises to earn profits. Wind and biofuels also enjoy preferred status under German law. As the world's sixth biggest producer of carbon dioxide emissions, Germany is trying to slash its output of greenhouse gases and wants renewable sources to supply a quarter of its energy needs by 2020 (Whitlock 2007). Germany also has decided to phase out all nuclear power plants by then.

Within months of its construction, the Geosol plant was forced to surrender its largest-in-the-world superlative once, twice, then six times, including the most recent (as of 2007) world's largest, the Solarpark Gut Erlasee in Bavaria, more than twice Geosol's size. On a former military base in Brandis, about twelve miles north of Espenhain, construction then began during 2007 on a solar array that will generate 40 megawatts, enough to power roughly ten thousand homes at German consumption rates, which are about half those of U.S. electrical customers (Whitlock 2007).

German companies that manufacture photovoltaic panels and other solar components employed forty thousand people in 2007. Another fifteen thousand worked for companies in solar-thermal firms, manufacturing heating systems. Matthias Machnik, an undersecretary in the German ministry of the environment, said that the country can't hope to compete with sun-blessed countries, but it can become a world leader in exports of solar technology and hardware. "Unless climate change accelerates, we only have a certain amount of available hours of sunshine," Machnik said. "For us, of course we will use solar power, but it is more important to secure the know-how for research and development" (Whitlock 2007).

During 2006, German exports comprised 15 percent of worldwide sales of solar panels and other photovoltaic equipment, according to industry officials. German companies hope to double their share of the global market, which amounted to $9.5 billion in 2006, growing by about 20 percent annually, said Carsten Koernig, managing director of the German Solar Industry Association (Whitlock 2007).

"It's been very important to create the necessary market in Germany," Koernig said. "We not only want to master the German market, but to conquer the world market as well." For now, the technology remains expensive and barely registers as a fraction of total energy production—less than 0.5 percent. The government hopes to increase that figure to 3 percent by 2020. (Whitlock 2007)

In northwestern Germany, Freiburg has been working to become the world's first "solar city," with a "solar-powered train station, energy-efficient row houses, innumerable rooftop photovoltaic systems and, high on a hill overlooking the vineyards, the world-famous Heliotrop, a high-tech cylindrical house that rotates to follow the sun" (Roberts 2004, 192). Freiberg is home to the Fraunhofer Institute for Solar Energy Systems, where scientists have been seeking breakthrough research that will reduce the cost of photovoltaic solar energy to competitive levels with fossil fuels and nuclear power, including a "multilayer" photovoltaic cell with an efficiency of 40 percent, twice present levels. Germany and Japan have implemented tax subsidies for rooftop solar energy systems. Given trends in research and cost reduction, solar energy in regions with abundant sunshine (such as the Middle East, Mediterranean, and U.S. Southwest) may be cost-effective before 2010.

## SOLAR POWER A PRIORITY IN PORTUGAL

Another "word's largest" was claimed in 2007 by Portugal, as a new solar array spread across 150 hilly acres near Serpa, about 120 miles southeast of Lisbon. This project was photovoltaic, constructed by General Electric Energy Financial Services and the PowerLight Corporation of the United States in partnership with the Portuguese company Catavento. "This is the most productive [largest capacity] solar plant in the world," its backers asserted. "It will produce 40 percent more energy than the second-largest one, Gut Erlasse in Germany," said Howard Wenger, speaking for PowerLight.

Unlike frequently misty Germany, southern Portugal is among the sunniest places in Europe, with as much as 3,300 hours of sunlight annually, nine hours during an average day. The Serpa solar array, when finished, will produce enough power to supply eight thousand homes. It will replace 30,000 tons' worth of annual greenhouse-gas emissions (Associated Press 2007g). This plant's photovoltaic system uses silicon solar cell technology to convert sunlight directly into electricity, producing 20 megawatts per year. "This project is successful because Portugal's sunshine is plentiful, the solar power technology is proven, government policies are supportive, and we are investing ... to help our customers meet their environmental challenges," said Kevin Walsh, managing director and leader of renewable energy at GE Energy Financial Services (Associated Press 2007g).

Portugal, which depends almost entirely on imported energy, also has been developing wave and solar power projects, as well as wind farms, planning to supply about 750,000 homes with nonfossil fuel energy before 2010. Portugal also has been exploring new hydropower projects and plans to invest €8 billion (about $11.5 billion) in renewable energy projects within five years. Socialist prime minister Jose Socrates anticipated during January 2007 that 45 percent of Portugal's energy will come from renewable sources by 2010 (Associated Press 2007g).

## SOLAR PAINT?

In the realm of imagination, envision solar paint—a layer of it on a car, or a building's outside wall, generating electricity. Chemical engineer Cyrus Wadia says, "Today this is science fiction; but everything we do is moving us toward that" (Environment News Service 2007x). The technique will be so simple, he says, that "anyone who feels comfortable in a kitchen could do it." A doctoral student with the University of California at Berkeley's interdisciplinary Energy and Resources Group, Wadia seeks solar paint by "synthesizing super-small nanoparticles." These particles, less than a billionth of a meter in diameter, are then suspended in solution. Wadia coats his solution on glass and analyzes his new device for "photocurrent," the current that flows through a photosensitive device as the result of exposure to radiant power (Environment News Service 2007x). Through his nanotech experiments, Wadia hopes to identify a material that is "extremely cheap, non-toxic, and abundant" in nature and suitable for manufacturing photovoltaic cells. Such a material "may not exist," he admits, "but we have to try" (Environment News Service 2007x).

## REFERENCES

Abboud, Leila. 2006. "Sun Reigns on Spain's Plains: Madrid Leads a Global Push to Capitalize on New Solar-Power Technologies." *Wall Street Journal*, December 5, A-4.

Associated Press. 2007g. "Portugal Celebrates Massive Solar Plant." *New York Times*, March 28. http://www.nytimes.com/aponline/technology/AP-Portugal-Solar-Power-Plant.html.

Auchard, Eric, and Leonard Anderson. 2006. "Google Plans Largest U.S. Solar-Powered Office." Reuters, *Washington Post*, October 16. http://www.washingtonpost.com/wp-dyn/content/article/2006/10/16/AR2006101601100_pf.html.

Dicum, Gregory. 2007. "Plugging into the Sun." *New York Times*, January 4. http://www.nytimes.com/2007/01/04/garden/04solar.html.

Environment News Service. 2007n. "First E.U. Commercial Concentrating Solar Power Tower Opens in Spain." March 30. http://www.ens-newswire.com/ens/mar2007/2007-03-30-02.asp.

———. 2007o. "Fresno Airport Goes Solar in a Big Way." April 13. http://www.ens-newswire.com/ens/apr2007/2007-04-13-09.asp#anchor4.

———. 2007x. "Solar Paint and Other Solar Surprises." November 12. http://www.ens-newswire.com/ens/nov2007/2007-11-12-094.asp.

Kaplow, Larry. 2001. "Solar Water Heaters: Israel Sets Standard for Energy; Cutting Dependence: Jerusalem's Alternative Energy Use a Lesson for United States." *Atlanta Journal and Constitution*, August 5, 1-P.

Kintisch, Eli. 2007a. "Light-Splitting Trick Squeezes More Electricity Out of Sun's Rays." *Science* 317 (August 3): 583–84.

Landauer, Robert. 2002. "Big Changes in Our China Suburb." *Sunday Oregonian*, October 20, F-4.

Mallaby, Sebastian. 2007. "Carbon Policy That Works: Avoiding the Pitfalls of Kyoto Cap-and-Trade." *Washington Post*, July 23, A-17. http://www.washingtonpost.com/wp-dyn/content/article/2007/07/22/AR2007072200884_pf.html.

Mufson, Steven. 2006. "A Sunnier Forecast for Solar Energy; Still Small, Industry Adds Capacity and Jobs to Compete with Utilities." *Washington Post*, November 20, D-1. http://www.washingtonpost.com/wp-dyn/content/article/2006/11/19/AR2006111900688_pf.html.

O'Connell, Sanjida. 2002. "Power to the People." *Times* (London), May 20, n.p. (in LEXIS).

Oppenheimer, Michael, and Robert H. Boyle. 1990. *Dead Heat: The Race against the Greenhouse Effect*. New York: Basic Books.

Revkin, Andrew C., and Matthew L. Wald. 2007. "Solar Power Captures Imagination, Not Money." *New York Times*, July 16. http://www.nytimes.com/2007/07/16/business/16solar.html.

Roberts, Paul. 2004. *The End of Oil: The Edge of a Perilous New World*. Boston: Houghton-Mifflin.

Rogers, Paul. 2006. "Solar Energy Heats Up." *Omaha World-Herald*, October 15, 1-RE, 2-RE.

Steinman, David. 2007. *Safe Trip to Eden: 10 Steps to Save Planet Earth from Global Warming Meltdown*. New York: Thunder's Mouth Press.

Wald, Matthew. 2007. "What's So Bad about Big?" *New York Times*, March 7. http://www.nytimes.com/2007/03/07/business/businessspecial2/07big.html.

Whitlock, Craig. 2007. "Cloudy Germany a Powerhouse in Solar Energy." *Washington Post*, May 5, A-1. http://www.washingtonpost.com/wp-dyn/content/article/2007/05/04/AR2007050402466_pf.html.

# CHAPTER 7

# Old Wine in New Bottles: Nuclear Power and Changes in Land Use

Energy planning with an eye to reduction of greenhouse gases has breathed new life into a number of old ideas, some of which, in earlier times, didn't get much play as "clean" energy. Witness nuclear fission, with its cargo of radioactive spent fuel. Reactors produce energy with no greenhouse emissions, but many environmentalists wouldn't be caught dead advocating more of them. Others, however, see nuclear as a necessary "bridge" source between the fossil-fueled past and a renewable-fueled future of solar, wind, and clean-source hydrogen. China has made a major commitment in this direction, hoping to substitute nuclear power partially for overreliance on low-energy "dirty" coal, which the Chinese have in abundance—so much, in fact, that China has passed the United States as the world's single largest national source of greenhouse emissions.

Other prosaic ideas can be remarkably good ways to shave greenhouse emissions. Farmers, for example, have become very good at no-till agriculture, encouraged by a new "green" attitude at the U.S. Department of Agriculture (USDA). The greening of urban areas—trees at streetside, gardens on rooftops—may shave several degrees off the thermometer in large cities, countering the urban heat-island effect and reducing the need for energy-gobbling air conditioning. Widespread planting of trees has widely been touted as a key strategy for countering warming, but the "cure" may not be that simple. How and where trees are planted can play a major role in whether reforestation helps reduce greenhouse emissions—it may, in some cases, actually make global warming worse. A dark forest, for example, will absorb more heat than arctic tundra covered with snow. At times, a thick field of grass can remove more carbon dioxide from the atmosphere than newly planted trees.

In some circumstances, such as along parts of England's eastern shore, authorities have decided to deal with a rising North Sea and subsiding coastline by giving land up to the sea. The bureaucratic euphemism is "managed realignment." The central government of Britain has even considered the ultimate in urban surrender, moving its capital out of London to higher ground.

## NUCLEAR POWER AS CLEAN ENERGY

James Lovelock, who pioneered measurement of trace gases in the atmosphere and developed the Gaia hypothesis, has become a staunch advocate of nuclear power to bridge the "gap" between fossil fuels and other sources of power. Gaia, named after the Greek goddess of the Earth, has been defined by Lovelock as a view that the planet acts as a living organism to maintain "life on Earth [that] actively keeps the surface conditions always favorable for whatever is the contemporary ensemble of organisms" (Volk 2006, 869). Lovelock, faced with scientific criticism, has since reformulated his school of thought as more abstract theory. He now asserts that human manipulation of greenhouse-gas levels in the atmosphere has stirred Gaia to declare war on humanity in which she "now threatens us with the ultimate penalty of extinction" (Volk 2006, 869). Such language strikes many other scientists as metaphorical and anthropomorphic. Pressed, Lovelock has agreed that the idea is a metaphor, with limited literal meaning (Volk 2006, 869).

In *The Revenge of Gaia*, Lovelock (2006) asserts that solar, wind, or biomass will take too much room (a quarter-million wind turbines, for example, to provide the power needs of the United Kingdom), and that

**Figure 7.1. James Lovelock in front of a statue of Gaia, 2000. © Photo Bruno Comby/IBC - Institut Bruno Comby/www.comby.org.**

nuclear fission should be used as a bridge to nuclear fusion and more efficient renewable sources of energy. He sees the risks of exposure to nuclear
waste as a small price to pay for its value as a carbon-free, proven source
of power. Fears of radiation are overblown, Lovelock contends. (Lovelock
also advocates large-scale geo-engineering solutions, such as sun-blocking
reflectors in space, when and if warming goes into a feedback-powered
runaway mode.) Tyler Volk, reviewing Lovelock's book in *Nature*, concluded: "Read this book for its thoughtful sections on global energy and
climate, but steer clear of its web of Old Testament-like prophecy" (Volk
2006, 870).

Barry Commoner, one of the founders of the modern environmental
movement, opposes Lovelock's position on nuclear power unconditionally.
Asked his view, Commoner replied:

> This is a good example of shortsighted environmentalism. It superficially
> makes sense to say, "Here's a way of producing energy without carbon diox-
> ide." But every activity that increases the amount of radioactivity to which
> we are exposed is idiotic. There has to be a life-and-death reason to do it. I
> mean, we haven't solved the problem of waste yet. We still have used fuel
> sitting all over the place. I think the fact that some people who have estab-
> lished a reputation as environmentalists have adopted this is appalling. (Vin-
> ciguerra 2007)

By 2007, China had begun construction on dozens of new nuclear
power plants to address part of its growing global-warming burden from
low-quality coal. China plans to spend $50 billion to build thirty-two nu-
clear plants by 2020. Some experts believe that China may build three
hundred more such plants by about 2050 to power what will be, by then,
the largest national economy in the world. By that time, China may have
half the nuclear-power capacity in the world (Eunjung Cha 2007, D-1).
China also is building the world's largest repository for spent nuclear fuel
in its western desert, amidst the Beishan Mountains, an area that is nearly
bereft of human habitation. The nuclear construction binge represents a
major change for China, where only 2.3 percent of electricity was generated from nuclear power in 2006, compared with about 20 percent in the
United States and almost 80 percent in France. In 2007, China had only
nine nuclear power plants.

In part because of Chinese demand, the price of processed uranium ore
increased from $10 to $120 a pound between 2003 and 2007. Expecting
that China will be one of its major customers, Japan's Toshiba paid $5.4
billion during 2006 to acquire a U.S. company, Westinghouse Electric,
which specializes in construction of nuclear plants. The Chinese government, emphasizing safety, has been using companies such as Westinghouse
to instruct its engineers who will build and operate new nuclear plants.

During the 1970s, energy interests in Sweden pitched nuclear power as
an antidote to oil. Seven new nuclear reactors came into operation
between 1972 and 1980. Because of environmental considerations, high
production costs, and low world market prices, Sweden's substantial uranium reserves (250,000 to 300,000 tons, or about 20 percent of known

world reserves) have not been exploited. By 2006, 45 percent of Sweden's electricity still was being generated by nuclear power, and 8 percent from fossil fuels.

Swedish nuclear power sustained a new blow during 2006 because of a near-meltdown at one of the country's reactors. An electricity failure on July 25, 2006, led to the immediate shutdown of the Forsmark 1 reactor after two of four backup generators, which supply power to the reactor's cooling system, malfunctioned for about twenty minutes.

This near-accident revived memories of the Ukraine's Chernobyl, which showered radioactive fallout over much of Sweden during April 1986. By 2010, all twelve of Sweden's nuclear plants will be shut down, to be replaced by natural gas, a fossil fuel, until other sources are available.

The United States also has been experiencing a wave of proposals for new nuclear energy capacity. Between 2007 and 2009, the Nuclear Regulatory Commission anticipates applications for building as many as thirty-two new nuclear reactors. The federal Energy Policy Act of 2005 included tax credits (up to $125 million over eight years), loan guarantees (as much as 80 percent of a plant's cost), and other benefits for nuclear power plant construction (Mufson 2007d, A-1). Localities sometimes add other incentives as well, including tax breaks and limits on liability.

## FARMING TECHNOLOGY IMPROVEMENTS

Contributions of farming to carbon dioxide in the atmosphere have been increasing with rising populations. Rattan Lal, a professor of soil science at Ohio State University, has asserted that the atmosphere's load of carbon dioxide could be reduced substantially by several relatively simple changes in farming technology. Carbon dioxide is added to the atmosphere via plowing, so Lal believes that reducing the depth of furrows will significantly reduce the amount of $CO_2$ introduced into the atmosphere by agriculture.

Farming with an eye to carbon sequestration utilizes soil restoration and woodland regeneration, no-till farming, cover crops, nutrient management, manuring and sludge application, improved grazing, water conservation, efficient irrigation, agroforestry practices, and growth of energy crops on spare lands. Intensive use of such practices, according to one estimate, could "offset fossil-fuel emissions by 0.4 to 1.2 gigatons of carbon per year, or 5 to 15 percent of the global fossil-fuel emissions" (Lal 2004, 1623).

During mid-2003, the USDA announced plans to give incentives to farmers for management practices that keep carbon in the soil. For the first time, the USDA began to factor reduction of greenhouse-gas emissions into soil conservation programs by giving priority to farmers who reduce emissions of carbon dioxide, methane, and nitrous oxides. Such programs represented $3.9 billion in federal spending during the 2003–2004 fiscal year (Clayton 2003, D-1). Farmers are being encouraged to use no- or low-tillage methods, as well as crop rotation, buffer strips, and other practices that reduce greenhouse-gas emissions and soil erosion.

Such practices are expected to retain 12 million tons of greenhouse gases in the soil by 2012 (Clayton 2003, D-1).

Tim O'Riordan of the Zuckerman Institute for Connective Environmental Research says:

> We have to put sustainable development at the heart of businesses such as fish farming and agriculture. We need agricultural stewardship schemes that have incentives for farmers to produce according to sustainable principles, which in turn will deliver healthy soil, water and wildlife. This, in turn, should offer jobs in recreation and education for eco-care. We also need the involvement of the local community to ensure that all acts of stewardship have neighborhood understanding and support. (Urquhart and Gilchrist 2002, 9)

On another front, work is under way on a genetically engineered rice seed that sharply reduces the crop's need for nitrogen fertilizer and thus its production of greenhouse gases. Eric Rey, president and chief executive officer of Arcadia Biosciences of Davis, California, is developing the seed and wants to market it in China so that carbon credits can be sold in global markets (Etter 2007a, A-1).

## COMPLICATIONS OF REFORESTATION

Proposals have been made to ameliorate increases in greenhouse-gas emissions through reforestation, the purposeful planting of large forests to absorb some of humankind's surplus carbon dioxide. The 1997 Kyoto Protocol contains mechanisms allowing governments of countries that produce more greenhouse gases per capita than average, such as the United States, to earn credit toward meeting their emissions goals by subsidizing the preservation of forests in poorer nations.

The problem with reforestation will be finding large tracts of land fertile enough to support trees that aren't already being used by human beings for other purposes. Reforestation might significantly slow greenhouse warming, but only if trees are planted on a very large scale. For example, if an area of 1.15 billion acres was planted, the trees on this land area, once mature, would remove almost 3 billion tons of carbon dioxide from the atmosphere per year, or about a third of the carbon that human beings add to the atmosphere. However, the creation of such a carbon sink would require a land area roughly half the size of the United States (Silver and DeFries 1990, 122–23).

In England, some private firms have been planting trees to offset their contributions to global warming. The London *Sunday Independent* conducted a campaign during which readers bought more than seven thousand trees to offset the amount of carbon dioxide created by the manufacture of the newspaper over a year's time. The newspaper itself contributed 750 trees. The Glastonbury arts festival sold 1,333 trees to offset the equivalent amount of carbon dioxide to all the emissions created in the setup, running, and dismantling of the show (Rowe 2000, 5). The trade in trees is coordinated by an organization called Future Forests. Some musical artists, such as the Pet Shops Boys and Neneh Cherry, produced 1.5 million "carbon-neutral" compact discs, meaning they have bought enough trees to offset the carbon emitted by production of their recording as well as movement of their stage materials.

## ARE FORESTS OVERRATED AS CARBON SINKS?

Increasing forest cover has been touted as a commonsense antidote to global warming. Conventional wisdom assumes that trees, because of their size, are the most voracious consumers of carbon dioxide among Earth's flora. The Kyoto Protocol is shot through with this assumption—which may not always be true. According to some scientific studies, lush grasslands under some conditions remove more carbon dioxide from the atmosphere than forests. Economist William R. Cline has pointed out that reforestation is "a temporary remedy because a forest stores additional carbon only when it is expanding; once it reaches a steady state, the carbon released by dying trees often offsets that sequestered by new and growing trees" (Cline 1992, 216–17).

Carbon uptake from reforestation has been proposed to reduce net carbon dioxide emissions to the atmosphere, even as some models indicate that forests sometimes contribute to global warming, on balance. For example, "The albedo [reflectivity] of a forested landscape is generally lower than that of cultivated land, especially when snow is lying. . . . In many boreal forest areas, the positive forcing induced by decreases in albedo can offset the negative forcing that is expected from carbon sequestration" (Betts 2000, 187). Stated briefly: forests, which are darker than snow-covered tundra, absorb more of the sun's heat. According to this analysis, high-latitude reforestation efforts actually might worsen the greenhouse effect.

Previous estimates of the amount of carbon stored by trees and shrubs may have been too high. According to an account by Cat Lazaroff for the Environment News Service, "This research could force climate experts to recalculate the benefits of growing trees as a way to offset human caused emissions of carbon dioxide" (Lazaroff 2002). Writing in *Nature*, Duke University ecologist Robert B. Jackson and his colleagues concluded that in many locations, trees may be absorbing less carbon than grass-covered soil. "It had been proposed that the woody species might even increase soil carbon compared to the grasslands," explained Jackson, lead author of the *Nature* study. "People really didn't think that grasslands would store more carbon in the soil than woodlands" (Lazaroff 2002). The rich black soils beneath many grasslands provide a long-term carbon repository. Jackson and colleagues explained:

> We found a clear negative relationship between precipitation and changes in soil organic carbon and nitrogen content when grasslands were invaded by woody vegetation, with drier sites gaining, and wetter sites losing, soil organic carbon. Losses of soil organic carbon at the wetter sites were

substantial enough to offset increases in plant biomass carbon, suggesting that current land-based assessments may overestimate carbon sinks. Assessments relying on carbon stored from woody plant invasions to balance emissions may therefore be incorrect. (Jackson et al. 2002, 623)

In general, the wetter a grassland, the greater its ability to remove carbon compared to trees. This study indicates, according to Jackson, that "as you move to increasingly wet environments, grasslands have a lot more soil carbon than shrublands and woodlands do. That was somewhat of a surprise. The analysis suggested that sites with the potential to store the most plant carbon also had the potential to lose the most soil organic carbon" (Lazaroff 2002).

In a "News and Views" article accompanying the study published in the same issue of *Nature*, Christine L. Goodale and Eric A. Davidson of the Woods Hole Research Center in Massachusetts noted that the work of Jackson and colleagues makes measurement of carbon sinks' effects on climate change much more complicated. "Woodlands, savannas, shrublands and grasslands cover about 40 percent of the Earth's land surface, and so their potential role as carbon sinks ... is a key factor in the global carbon budget," they wrote. "Measuring the effects of woody encroachment at particular sites is one challenge; extrapolating the results to regional or larger scales is quite another. Particular sites are certainly large sinks for carbon, but the global extent of grassland replacement by shrubland is highly uncertain" (Goodale and Davidson 2002, 594).

Jackson and colleagues received support from another study indicating that agricultural lands may sequester more atmospheric carbon dioxide than forests in rivers. This study was conducted by researchers at the Yale School of Forestry and Environmental Studies and the Institute for Ecosystem Studies in Millbrook, New York. Carbon dioxide dissolved in rain and water in the soil acts as an acid, reacting with subterranean rocks to form dissolved carbonate alkalinity, which is then transported to the ocean. Peter Raymond, assistant professor of ecosystem ecology at Yale, and Jonathan Cole, an aquatic biologist at the Institute of Ecosystem Science, asserted that dissolved carbonate alkalinity emanating from the Mississippi River has increased dramatically during the past half-century. They argued that the increase in dissolved alkalinity export is related to increases in precipitation that they document in the Mississippi watershed (Raymond and Cole 2003).

This research contradicts assumptions that converting agricultural fields to forests increases removal of atmospheric carbon dioxide by locking it into trees and soils. "Chemical weathering and the subsequent export of carbonate alkalinity from soils to rivers account for significant amounts of terrestrially sequestered carbon dioxide," Raymond and Cole wrote (2003, 88). They found that increases in alkalinity export from the Mississippi River during the previous fifty years were related not only to higher rainfall but also to the amount and type of land cover. "These observations have important implications for the potential management of carbon sequestration in the United States," they stated (Raymond and Cole 2003, 88).

Other research also suggests that planting forests to curb global warming could backfire. Thus, planting trees may not do much to curb global warming, contrary to assumptions expressed in the Kyoto Protocol. This assumption is deeply flawed, according to Richard Betts of Britain's Meteorological Office. "Carbon accounting alone will overestimate the contribution of afforestation to reducing climate warming," he told *New Scientist* (2001). Betts presented detailed calculations showing that planting trees across the snow-covered swaths of Siberia and North America will heat the planet rather than cool it. On locations other than the tundra, the cooling potential of forests also is much less than previously supposed, he believes (*New Scientist* 2001). Where forests replace snowy tundra, which usually reflects large amounts of solar radiation, heating of the Earth could accelerate. Betts calculates that at northern latitudes, warming as a result of planting forests will overwhelm any cooling effect from the trees soaking up carbon dioxide.

"Canada and Russia have proposed to plant forests in their empty tundras to help meet their Kyoto commitments, because a hectare [2.5 acres] of immature forest can absorb more than 100 tons of carbon each year, despite growing slowly" (*New Scientist* 2001). Betts has calculated that the net warming effect of heat-absorbent forests in both regions is equivalent to an annual emission of 30 tons of carbon per acre. "I am not suggesting that we deforest," says Betts. "But afforestation is not always an effective alternative to cutting fossil fuel emissions" (*New Scientist* 2001).

William H. Schlesinger and John Lichter confirmed a view that forests have been overrated as a carbon sink:

> Such findings call into question the role of soils as long-term carbon sinks, and show the need for a better understanding of carbon cycling in forest soils. ... Fast turnover times of organic carbon in the litter layer (of about three years) appear to constrain the potential size of the carbon sink. Given the observation that carbon accumulation in the deeper mineral soil layers was absent, we suggest that significant, long-term net carbon sequestration in forest soils is unlikely. (Schlesinger and Lichter 2001, 464)

Other research indicates that older, wild forests are far better than plantations of young trees at removing carbon dioxide from the atmosphere. One such analysis, published in the journal *Science*, was completed by Dr. Ernst-Detlef Schulze, the director of the Max Planck Institute for Biogeochemistry in Jena, Germany, and two other scientists at the institute. The study provided an important new argument for protecting old-growth forests. The scientists said that their study provides a reminder that the main goal should be to reduce carbon dioxide emissions at the source: smokestacks and tailpipes.

"In old forests, huge amounts of carbon taken from the air are locked away not only in the tree trunks and branches, but also deep in the soil, where the carbon can stay for many centuries," said Kevin R. Gurney, a research scientist at Colorado State University. When such a forest is cut, he said, almost all of that stored carbon is eventually returned to the air in the form of carbon dioxide. "It took a huge amount of time to get that carbon sequestered in those soils," he said, "So if you release it, even if you plant again, it'll take equally long to get it back" (Revkin 2000b, A-23).

The German study, together with other similar research, has produced a picture of mature forests that differs sharply from long-held notions in forestry, Schulze said. He said that aging forests were long perceived to be in a state of decay that releases as much carbon dioxide as it captures. Soils in undisturbed tropical rain forests, Siberian woods, and some German national parks contain enormous amounts of carbon derived from fallen leaves, twigs, and buried roots that can bind to soil particles and remain stored for a thousand years or more. When such forests are cut, the trees' roots decay and soil is disrupted, releasing the carbon dioxide (Revkin 2000b, A-23; Schulze, Wirth, and Heimann 2000). "In contrast to the sink management proposed in the Kyoto Protocol, which favors young forest stands, we argue that preservation of natural old-growth forests may have a larger effect on the carbon cycle than promotion of re-growth," said Schulze and his colleagues (2000, 2058). Instead of reducing the level of carbon dioxide in the atmosphere, they explained, the Kyoto Protocol's emphasis on new growth at the expense of established forests

> will lead to massive carbon losses to the atmosphere mainly by replacing a large pool with a minute pool of re-growth and by reducing the flux into a permanent pool of soil organic matter. Both effects may override the antici-pated aim, namely to increase the terrestrial sink capacity by afforestation and reforestation. (Schulze, Wirth, and Heimann 2000, 2059)

## "MANAGED REALIGNMENT" ALONG ENGLAND'S COASTLINE

Parts of England's coastline are afflicted by the same problems as the U.S. East and Gulf of Mexico coasts. The land is subsiding, while icemelt and thermal expansion slowly raise sea levels. The U.K. Climate Impact Programme, a government-funded program at Oxford University, forecasts that the mean sea level (combining rising water and subsidence of land) could rise as much as a meter (3.3 feet) in this area by late in the twenty-first century. In addition to climate change, "isostatic rebound," the rise of Scotland's coast following the last ice age, is contributing to a subsiding coastline southward along the English coast. As the sea rises 3 millimeters (an eighth of an inch) a year abreast of Essex, the land itself is sinking half as rapidly. By 2003, the rising waters were threatening the closed Bradwell Nuclear Power Station (*Guardian* 2003, 4).

Anthony Browne commented in the London *Times*:

> A thousand years after King Canute showed that man could not hold back the tide, the Government has come to the same conclusion. The Environ-ment Agency, the government body responsible for flood defenses, is

planning a strategic withdrawal from large parts of the English coastline because it believes that it can no longer defend them from rising sea levels, the result of global warming. (Browne 2002, 8)

The new strategy, officially called "managed realignment," allows the sea to flood low-lying farmland, as the government abandons attempts to fend off invading seawaters by building ever-higher defenses. The policy, which will allow the encroaching sea to submerge several thousand acres of land, has been welcomed by environmentalists in part because it creates wetland habitats for endangered wildlife. Many farmers, however, have demanded more elaborate flood defenses in hopes that their land can be salvaged. The area of affected coastline ranges from the Humber Estuary, around East Anglia, to the Thames Estuary and west to the Solent. Strategic withdrawal also has been planned for sections of the Severn Estuary (Browne 2002, 8). The first site surrendered to the sea was in Lincolnshire, as roughly two hundred acres of farmland was flooded by seawater at Freiston Shore after flood defense banks were breached to create a saltmarsh bird reserve (Browne 2002, 8).

"Managed realignment is in its infancy . . . but it has been an emotive issue; a lot of people are concerned about it. But you have to get the message across that we are defending property," said Brian Empson, flood defense policy manager for the Environment Agency (Browne 2002, 8). The land that was returned to the sea in 2002 had been reclaimed 150 years earlier and protected by man-made banks. A grass-covered wall in Abbots Hall farm country on the east coast of Essex that had held back the sea for almost four hundred years also was breached intentionally during 2002. The use of breakwaters made of clay, bricks, and finally large blocks of concrete had not stopped rising waters.

Until recently, the Thames River barrier, built to protect London and surrounding areas from unusually high tides and storm surges, closed an average of two or three times a year. Between November 2001 and March 2002, however, the barrier was raised twenty-three times. A British report released in September 2002, said that 59,000 square miles (home to 750,000 people) in and around London are vulnerable to flooding because they are below high-tide levels.

## REFERENCES

Betts, Richard A. 2000. "Offset of the Potential Carbon Sink from Boreal Forestation by Decreases in Surface Albedo." *Nature* 408 (November 9): 187–90.

Browne, Anthony. 2002. "Canute Was Right! Time to Give Up the Coast." *Times* (London), October 11, 8.

Clayton, Chris. 2003. "U.S.D.A. Will Offer Incentives for Conserving Carbon in Soil." *Omaha World-Herald*, June 7, D-1, D-2.

Cline, William R. 1992. *The Economics of Global Warming.* Washington, D.C.: Institute for International Economics.

Etter, Lauren. 2007a. "In China, Plan to Turn Rice into Carbon Credits." *Wall Street Journal*, October 9, A-1, A-15.

Eunjung Cha, Ariana. 2007. "China Embraces Nuclear Future; Optimism Mixes with Concern as Dozens of Plants Go Up." *Washington Post*, May 29. http://www.washingtonpost.com/wp-dyn/content/article/2007/05/28/AR2007052801051_pf.html.

Goodale, Christine L., and Eric A. Davidson. 2002. "Carbon Cycle: Uncertain Sinks in the Shrubs." *Nature* 418 (August 8): 601.

*Guardian* (London). 2003. "Rising Tide: Who Needs Essex Anyway." June 12, 4.

Jackson, Robert B., Jay L. Banner, Esteban G. Jobbagy, William T. Pockman, and Diana H. Wall. 2002. "Ecosystem Carbon Loss with Woody Plant Invasion of Grasslands." *Nature* 418 (August 8): 623–26.

Lal, R. 2004. "Soil Carbon Sequestration Impacts on Global Climate Change and Food Security." *Science* 304 (June 11): 1623–27.

Lazaroff, Cat. 2002. "Replacing Grass with Trees May Release Carbon." Environment News Service, August 8. http://ens-news.com/ens/aug2002/2002-08-08-07.asp.

Lovelock, James. 2006. *The Revenge of Gaia: Why the Earth Is Fighting Back—and How We Can Still Save Humanity.* London: Allen Lane.

Mufson, Steven. 2007d. "Nuclear Power Primed for Comeback; Demand, Subsidies Spur U.S. Utilities." *Washington Post*, October 8, A-1.

*New Scientist.* 2001. "Planting Northern Forests Would Increase Global Warming." Press release, July 11. http://www.newscientist.com/news/news.jsp?id=ns99991003.

Raymond, Peter A., and Jonathan J. Cole. 2003. "Increase in the Export of Alkalinity from North America's Largest River." *Science* 301 (July 4): 88–91.

Revkin, Andrew C. 2000b. "Planting New Forests Can't Match Saving Old Ones in Cutting Greenhouse Gases, Study Finds." *New York Times*, September 22, A-23.

Rowe, Mark. 2000. "When the Music's Over . . . a Forest Will Rise." *Independent* (London), June 25, 5.

Schlesinger, W. H., and J. Lichter. 2001. "Limited Carbon Storage in Soil and Litter of Experimental Forest Plots under Increased Atmospheric $CO_2$." *Nature* 411 (May 24): 466–69.

Schulze, Ernst-Detlef, Christian Wirth, and Martin Heimann. 2000. "Managing Forests after Kyoto." *Science* 289 (September 22): 2058–59.

Silver, Cheryl Simon, and Ruth S. DeFries. 1990. *One Earth, One Future: Our Changing Global Environment.* Washington, D.C.: National Academy Press.

Urquhart, Frank, and Jim Gilchrist. 2002. "Air Travel to Blame as Well." *Scotsman*, October 8, n.p. (in LEXIS).

Vinciguerra, Thomas. 2007. "At 90, an Environmentalist from the '70s Still Has Hope." *New York Times*, June 19. http://www.nytimes.com/2007/06/19/science/earth/19conv.html.

Volk, Tyler. 2006. "Real Concerns, False Gods: Invoking a Wrathful Biosphere Won't Help Us Deal with the Problems of Climate Change" [review of James Lovelock, *The Revenge of Gaia*] *Nature* 440 (April 13): 869–70.

# CHAPTER 8

# The Political Economy
# of Global Warming

Suddenly, after Democrats came to power in both houses of the U.S. Congress during the November 2006 elections, the idea of legal limits on greenhouse gases (usually conjoined with a carbon trading scheme of some sort) became an idea whose time had come in corporate circles. Many executives decided it was better to be part of the decision-making process than to have limits imposed upon them. Within six months, even some oil and automobile companies supported the idea (at least in theory), as global warming became a major subject of news coverage and public debate, from sea to shining sea. Businesspeople of all stripes suddenly realized that trading schemes (not to mention a paradigm shift in energy manufacture and use) could be profitable.

Global warming suddenly emerged as an issue in corporate boardrooms. During April 2007, ConocoPhillips became the first major U.S. oil company to call for a federal carbon dioxide emissions cap. Even some Republicans took up the cry, through Republicans for Environmental Progress (www.repamerica.org), which advocates progressive environmental policies within the party. The Rockefeller Foundation at about the same time put up $70 million over five years to help Asian and African farmers and cities cope with floods, droughts, and other hazards of global warming, as the World Bank established a new set of standards for its third-world projects to make them "climate-proof."

Famous names and big money came to the fore with dizzying speed. Bill Clinton's foundation made global warming one of its priorities, as London-based HSBC (Hongkong and Shanghai Banking Corp.) on May 31, 2007, announced a five-year, $100 million partnership to combat the "urgent threat" of climate change worldwide, most notably in five large cities: Hong Kong, London, Mumbai, New York, and Shanghai. The

$100 million donation was to be shared with four partner organizations: the Climate Group, Earthwatch Institute, the Smithsonian Tropical Research Institute, and the World Wildlife Fund (WWF).

Leaders from 150 multinational companies published a full-page advertisement in London's *Financial Times* on November 29, 2007, advocating legally binding greenhouse-gas limits. The signers included some of the world's largest companies—Coca-Cola, General Electric, Shell, Nestlé, Nike, DuPont, Johnson & Johnson, British Airways, and Shanghai Electric, among others—saying that scientific evidence for global warming is "now overwhelming" and calling for an agreement worldwide that "will provide business with the certainty it needs to scale up global investment in low-carbon technologies" (Eilperin 2007b, A-3).

"You may find it surprising that someone from business would sit here and talk about regulations," said James Smith, chairman of Shell UK. Smith said that legal standards will be required to "give business the confidence to make those long-term investments in lower-carbon technologies" (Eilperin 2007b, A-3).

Green headlines were coming from unusual places. Stop the presses! Rupert Murdoch, high praiser of George W. Bush and one of the world's largest publishing (and media) magnates, pledged during May 2007 that his News Corporation would "reduce its carbon footprint to zero" within three years. "Climate change poses clear, catastrophic threats," Murdoch declared (Kurtz 2007, C-1).

## THEORY AND PRACTICE

With all the excitement over the new "green" attitude in the United States, however, many of the proposals found slow going in the Senate and the House. A realization began to emerge that, while politicians and corporate executives found talking green to be easy, acting in ways that might cost short-term profits and require sacrifices from consumers brought little cheer. We not only have a special-interest-driven system, but politicians won't lead even when a problem is obvious, if it may cost constituents money. "Democracy," such as we have it, may be the death knell of the world we know.

Lawmakers began to doubt that they could harvest voters by asking (or—heavens, no!—*requiring*) voters to abide by greenhouse-gas reductions that would cost them comfort, convenience, and money. Auto manufacturers flashed green credentials, but trenchantly opposed stiffer mileage standards. Old habits die hard. For all the hip-green facades put up by a few oil companies, the first hint that Congress might reduce their depletion allowances and other tax subsidies brought forth a cavalcade of television and full-color, full-page newspaper advertisements inveighing with mighty gravitas about how "energy tax hikes" would bring back the long gas lines of the 1970s.

As long as they were talking theory, environmental responsibility became a corporate and political mantra—no more foreboding of global warming solutions as a left-wing invitation to go hungry in a cold shower.

Climate change became "an urgent problem that requires global action," according to a joint statement issued by the leaders of more than ninety major international corporations and organizations, including Citigroup, General Electric, Rolls-Royce, Volvo, and the World Council of Churches. The statement called on governments to set new targets for reducing greenhouse-gas emissions and enact "bold policies to increase energy efficiency." The signatories recommended that "the world set a price on emissions of carbon dioxide ... and cooperate on a new international agreement to replace the Kyoto Protocol" (Environment News Service 2007g).

The sudden eagerness to participate in a sustainable future was palatable. "Cost-efficient technologies exist today, and others could be developed and deployed, to improve energy efficiency and to help reduce emissions of $CO_2$ and other greenhouse gases in major sectors of the global economy," the same statement said. "Research indicates that heading off the very dangerous risks associated with doubling pre-industrial atmospheric concentrations of $CO_2$, while an immense challenge, can be achieved at a reasonable cost" (Environment News Service 2007g).

The statement was an outcome of discussions by participants of the Global Roundtable on Climate Change, initiated during 2004 by the Earth Institute at Columbia University. "Leaders from key economic sectors and regions of the world have reached a consensus on the path forward to reduce human-made climate change," said Jeffrey Sachs, chair of the roundtable and director of the Earth Institute. "This initiative points the way to an urgently needed global framework for action" (Environment News Service 2007g).

Other signatories included Air France, American Electric Power, Bayer, China Renewable Energy Industry Association, Electricity Generating Authority of Thailand, ENDESA, Eni, Eskom, FPL Group, Iberdrola, ING, Interface, Marsh & McLennan Companies, Munich Re, NRG Energy, Patagonia, Ricoh, Stora Enso North America, Suntech Power, Swiss Re, Vattenfall, and the World Petroleum Council (Environment News Service 2007g).

The new green was popping up in corporate suites around the world. Sir Terry Leahy, the chief executive officer (CEO) of Tesco, a large British supermarket chain, announced that the company had budgeted almost $1 billion over five years to lead "a revolution in green consumption." Tesco proposed to affix labels to Tesco-branded products that tally their carbon footprints and to create a Sustainable Consumption Institute that will develop standards for measuring carbon that could be used throughout the food-retailing industry (Cortese 2007). A commonly accepted measure, Sir Terry said, "will enable us to label all our products so that customers can compare their carbon footprint as easily as they can currently compare their price or their nutritional profile" (Cortese 2007).

One major problem with all this shine-on green rhetoric was that while posturing raises an issue's salience, it will not, by itself, reduce greenhouse-gas emissions. "Politicians pander to 'green' constituents who want to feel good about themselves. Grandiose goals are declared. But measures to achieve them are deferred—or don't exist," wrote Robert J. Samuelson (2007) in the *Washington Post*.

Samuelson singled out California governor Arnold Schwarzenegger, whose state has passed a law declaring that greenhouse-gas emissions by 2020 will be cut to 1990 levels (about 25 percent). The state's legislative and executive branches have been debating proposals aiming for an 80 percent reduction below 1990 levels by 2050. "However," said Samuelson,

> the policies to reach these goals haven't yet been formulated; that task has been left to the California Air Resources Board [CARB]. Many mandates wouldn't take effect until 2012, presumably after Schwarzenegger has left office. As for the 2050 goal, it's like his movies: make-believe.... But it's respectable make-believe. Schwarzenegger made the covers of *Time* and *Newsweek*. The press laps this up; "green" is the new "yellow journalism," says media critic Jack Shafer. Naturally, there's a bandwagon effect. At least 35 states have "climate action plans." None of this will reduce global greenhouse-gas emissions from present levels. (2007, A-15)

While politicians posture, energy use rises, with only incremental gains in efficiency, and some construction of nonfossil fuel infrastructure. The McKinsey Global Institute anticipates that, given present trends, "worldwide energy use will have risen 45 percent from 2003 to 2020" (Samuelson 2007, A-15).

By late 2007, however, the CARB was beginning to move on specific regulations to cut greenhouse gases—such things as requiring ships in docks to shut off their engines and plug into landside electrical grids, trucks to be more aerodynamic, and tire pressure on cars to be checked when oil is changed. That is just the beginning. The CARB by the end of 2008 was applying its regulatory authority to transportation, power plants, cement manufacturers, and oil refineries, all aiming to cut emissions.

## THE U.S. CLIMATE ACTION PARTNERSHIP

By early in 2007, green credentials were all the rage in corporate America. In the bad old days when the official line in the U.S. Senate was that global warming was a hoax, and when "sound science" at the White House meant inviting science-fiction writer Michael Crichton over for a chat, one never would have seen this: the U.S. Climate Action Partnership (USCAP)—a coalition of chief executives leading several major corporations, including DuPont, General Electric, Xerox, Rio Tinto (a multinational mining conglomerate), and Duke Energy—arm-in-arm with leaders of four national environmental groups, all declaring their fidelity to mandatory controls that would reduce greenhouse gas emissions by 60 to 80 percent in fifty years (Pegg 2007a). By late 2007, USCAP had enrolled twenty-seven of the world's largest companies, with combined revenues of $2 trillion and 2.7 million employees (Environmental Defense 2007, 9).

How's this for a ringing declaration—and from the CEO of a coal-burning electric utility: "We share a view that climate change is the most pressing environmental issue of our time and we agree that as the world's

**Figure 8.1. Senator Barbara Boxer.**
Source: Boxer.senate.gov.

largest source of global warming emissions our country has an obligation to lead." Those were the words of Peter Darbee, chairman and CEO of PG&E, California's largest gas and electric utility (Pegg 2007a). The U.S. economy "is the world's locomotive," Darbee told a Senate panel, adding that the members of USCAP "believe it is critical to get the engine pulling in the right direction on climate change" (Pegg 2007a). Sen. Barbara Boxer, a California Democrat and chair of the environment committee, said the USCAP recommendations marked a "turning point" in the U.S. debate over controlling greenhouse gas emissions. "The companies and groups before us today also make clear that by acting now, we can help, not hurt our economy," she said (Pegg 2007a).

USCAP partners support six recommendations:

- Account for the global dimensions of climate change—U.S. leadership is essential for establishing an equitable and effective international policy framework for robust action on climate.
- Recognize the importance of technology—The cost-effective deployment of existing energy efficient technologies should be a priority.
- Be environmentally effective—Mandatory requirements and incentives must be stringent enough to achieve necessary emissions reductions.
- Create economic opportunity and advantage—A climate protection program must use the power of the market to establish clear targets and timeframes.
- Be fair—Solutions must account for the disproportionate impact of both global warming and emissions reductions on some economic sectors, geographic regions and income groups.

- Encourage early action—Prior to the effective date of mandatory pollution limits, every reasonable effort should be made to reduce emissions (Environment News Service 2007w).

Even as the green tide rolled in around him, Sen. James Inhofe, a Republican from oil-rich Oklahoma, who had once laid down the climatic law on the environment committee (and called global warming a "hoax") had not changed his tune. He just had no power in 2007, as the new committee chair, Senator Boxer, brought down the gavel and said, "Senator Inhofe, you used to do this. Now I do it. Please sit down." Inhofe was still sore, as he called the members of USCAP "climate profiteers," motivated, he said, by self-interest.

"More and more companies that wish to profit on backs of consumers are coming out of the woodwork to endorse climate proposals in hopes of forcing customers to buy their products or to penalize their competitors," Inhofe said (Pegg 2007a). Oil companies' profits—Exxon had just reported the largest profits in the history of capitalism—don't bother Inhofe at all. The senator complained that the companies involved in USCAP, such as DuPont and BP, had invested heavily in renewable energy technologies. He sounded a lot like a typewriter salesman at a computer convention.

Even as Inhofe blustered, oil companies that once agreed with his position made plans to reduce their carbon footprints. Royal Dutch Shell CEO Jeroen van der Veer displayed that consensus in an introduction to the company's *Sustainability Report 2006*:

> I have said repeatedly that, for us, the debate about $CO_2$'s impact on the climate is over. I am pleased at how our people are responding to my call to

**Figure 8.2. Senator James Inhofe.**
Source: Inhofe.senate.gov.

find ways to mitigate $CO_2$ impacts from fossil fuels. Our focus is on what we can do to reduce $CO_2$ emissions. We are determined to find better, lower-cost ways to capture and store $CO_2$. (Environment News Service 2007w)

ConocoPhillips's chairman and CEO Jim Mulva agreed:

> We recognize that human activity, including the burning of fossil fuels, is contributing to increased concentrations of greenhouse gases in the atmosphere that can lead to adverse changes in global climate. ... While we believe no one entity alone can address the environmental, economic and technological issues inherent in any solution, ConocoPhillips will show leadership in finding pragmatic and sustainable solutions. (Environment News Service 2007w)

Mulva said that ConocoPhillips was beginning to factor the long-term costs of excess carbon dioxide into capital spending plans for its major projects, finding ways to improve energy efficiency, including a specific commitment to a 10 percent improvement in energy efficiency at its U.S. refineries by 2012 (Environment News Service 2007w).

## A POLITICAL QUESTION: IS CARBON DIOXIDE A POLLUTANT?

Anything, in excess, is a problem. Too much chocolate is toxic. Too many potatoes, eaten at once, can be poisonous. Too much sunshine can provoke skin cancer, although a certain amount is necessary for life as we know it. Likewise, carbon dioxide: We need some of it to prevent the Earth from freezing, but too much, and we will fry. Witness Venus, with an atmosphere that is 95 percent carbon dioxide, a victim of a runaway greenhouse effect.

The U.S. Supreme Court recognized this logic early in 2007 when it held, 5 to 4, in *Massachusetts v. Environmental Protection Agency*, against the express wishes of the George W. Bush administration, that the federal government has the power to regulate carbon dioxide and other greenhouse gases from vehicles. Greenhouse gases are air pollutants under the landmark environmental law, Justice John Paul Stevens said in his majority opinion. The court's four conservative justices—Chief Justice John Roberts and justices Samuel Alito, Antonin Scalia, and Clarence Thomas— dissented (Associated Press 2007c).

The Court addressed three issues: First, do states have the right to sue the Environmental Protection Agency (EPA) to challenge its decision? Second, does the Clean Air Act give the EPA the authority to regulate tailpipe emissions of greenhouse gases? Third, does the EPA have the discretion *not* to regulate those emissions? The majority affirmed the first two questions. On the third, it ordered EPA to reevaluate its assertion that it has the discretion not to regulate tailpipe emissions, which the Court found to be out of alignment with the Clean Air Act.

As it often does, the Supreme Court seemed to be taking stock of a changing political consensus in the country at a time when Congress and even some oil companies (and later in the year, even Bush himself) were paying at least lip service to the dangers of global warming. "In many ways, the debate has moved beyond this," said Chris Miller, director of the global warming campaign for Greenpeace, one of the environmental groups that sued the EPA. "All the front-runners in the 2008 presidential campaign, both Democrats and Republicans, even the business community, are much further along on this than the Bush administration is" (Associated Press 2007c).

Republican governors Schwarzenegger of California and Jodi Rell of Connecticut criticized President Bush's lack of reaction to the Supreme Court ruling:

> California, Connecticut, and 10 other states are poised to enact tailpipe emissions standards—tougher than existing federal requirements—that would cut greenhouse gas emissions from cars, light trucks and sports-utility vehicles by 392 million metric tons by the year 2020, the equivalent to taking 74 million of today's cars off the road for an entire year.
>
> Since transportation accounts for one-third of America's greenhouse gas emissions, enacting these standards would be a huge step forward in our efforts to clean the environment and would show the rest of the world that our nation is serious about fighting global warming....
>
> Under the Clean Air Act, California has the right to enact its own air pollution standards, which other states may then follow, as long as the EPA grants California a waiver. The waiver gives California, and other states, formal permission to deviate from federal standards. California has requested more than 40 such waivers over the past 30 years and has been granted full or partial permission for most of them.
>
> By continuing to stonewall California's request, the federal government is blocking the will of tens of millions of people in California, Connecticut and other states who want their government to take real action on global warming. (Schwarzenegger and Rell 2007, A-13)

On September 14, 2007, a federal court in Vermont sided with the states that had adopted new clean-car standards enacted by California in a decision that opened a legal avenue for new limits on greenhouse-gas emissions from automobiles. These may reduce global warming emissions from cars about 30 percent when they are fully effective in 2016.

## GREAT BRITAIN LEADS

Great Britain, where even the Conservatives favor curtailing carbon emissions significantly, has long been a world leader in greenhouse-gas diplomacy. During mid-March 2007, its government became the first to propose binding laws enforcing a 60 percent reduction in greenhouse-gas emissions by 2050; some British environmental groups favored an 80 percent reduction. The British proposal came within a week of a European Union proposal to cut carbon emissions 20 percent by 2020. Britain exceeded that target, committing to a 26 to 32 percent reduction during the same period (Cowell 2007).

The "Climate Change Bill," as it was known, could change daily life in Britain. A *New York Times* analysis anticipated that

officials might be summoned to appear before judges for failing to meet targets, households could be pressed to switch to low-energy light bulbs and install home insulation, and manufacturers could be asked to build TV sets without standby modes that consume energy when the devices are not in use. (Cowell 2007)

Consumers' use of autos and airline transport may be inhibited as well. One proposal would heavily tax air travel after a person's first flight per year. "This bill is an international landmark," the environment minister, David Miliband, told reporters. "It is the first time any country has set itself legally binding carbon targets. It is an environmental contract for future generations" (Cowell 2007).

Addressing climate change has been an influential political issue in Britain for many years. Even Conservative prime minister Margaret Thatcher, a former chemistry teacher, embraced it during the 1980s. In 2007, Prime Minister Tony Blair called global warming "the biggest long-term threat facing our world" (Cowell 2007). Blair called for a revolution in how Britons drive, heat their homes, run their businesses, and schedule vacation flights. Nearly every step that government, corporations, customers, and constituents took would be measured against carbon-generation standards. The draft law would require a carbon-trading system and five-year "carbon budgets" anticipated fifteen years in advance.

The various parties in Britain competed with each other for the most effective plans to address global warming, a central issue in national elections, likely to be fought by the Conservative leader, David Cameron, and Gordon Brown, chancellor of the exchequer, who assumed the prime minister's job after Blair stepped down during the summer of 2007. Cameron even advocated new taxes on airline travel, which would be a matter of far-left advocacy in the United States (Cowell 2007).

Even so, some commentators in Britain wrung their hands over the ascension of green consciousness in Britain, asserting that the country was too small to have an important, practical role in the worldwide climatic equation. In the March 13, 2007, London *Evening Standard*, columnist Nirpal Dhaliwal said there was "more than a whiff of colonial condescension about British politicians' attitudes to developing world industrialization." He said Britain's share of global carbon emissions was very low—around 2 percent—while China was building a new coal-burning power station every two weeks. "We could decide to live in the Stone Age burning nothing, and it would have virtually no impact on the overall problem of global warming," he wrote. (Cowell 2007)

## CAP-AND-TRADE SYSTEMS

A cap-and-trade system combines a limit on carbon emissions with a market rewarding industries that reduce their production of carbon dioxide and other greenhouse emissions. This type of system allows emitters to purchase emissions "permits" from businesses that reduce their emissions below allocation levels (Chameides and Oppenheimer 2007, 1670).

The cap-and-trade system originated in the United States as a program that reduced sulfur dioxide emissions from acid rain. Lead was eliminated from gasoline, and ozone-depleting chemicals (principally chlorofluorocarbons or CFCs) were phased out with similar economic incentives. A greenhouse-gas cap-and-trade system in the United States could be linked with world markets as established by the Kyoto Protocol. An international cap-and-trade system also could be used to reward nations that have tropical forests for reducing deforestation, meanwhile preserving irreplaceable habitat. Ensuring the integrity of such a system will require rigorous monitoring and auditing (Chameides and Oppenheimer 2007, 1670).

Indications have been strong that the United States will rejoin the world on climate change after George W. Bush's second term ends. The new Democratic majorities in both houses of Congress have been considering cap-and-trade systems, and even major Republican presidential candidates in 2008 (an example being John McCain) have spoken out on the necessity of curtailing greenhouse-gas emissions.

By mid-2007, having negotiated with lobbyists for unions and industries (including some electric utilities), the Senate was moving toward passage of the Low Carbon Economy Act, a bill sponsored by senators Jeff Bingaman (D-N.Mex.) and Arlen Specter (R-Pa.) to establish a cap-and-trade on emissions of greenhouse gases. The measure also limits the price that industries would pay for permits to emit greenhouse gases. The bill also contains billions of dollars to help Alaska cope with the extraordinary effects of warming there (Broder 2007). Some environmental groups asserted that the measure was half-hearted. The Bush White House is opposed to any cap-and-trade plan. The proposal sets target emissions for 2020 at 2006 levels and for 2030 at 1990 levels.

Under the plan as proposed, emissions permits in excess of those granted to companies could be purchased for $12 per metric ton of carbon dioxide emissions in the first year, rising by 5 percent above the rate of inflation each year after that. The proceeds would be allocated to finance research into clean energy, to counter "effects of global warming, compensate farmers for higher fuel costs and help low-income families pay their heating and gasoline bills" (Broder 2007). "A good sign of the legislation's limits is the fact that it's backed by some of our country's biggest coal companies and polluters," Brent Blackwelder, president of Friends of the Earth, said in a statement. "This legislation is flawed in many ways." He said that the bill would actually allow global warming emissions to increase at first and that the bill's caps would not bring emissions back to 2006 levels until 2020.

"This is a clear attempt to find the political middle ground on the climate debate rather than a utopian solution," said Frank O'Donnell, president of Clean Air Watch. But he added that "the bill appears to fall well short of what science says is needed." (Mufson 2007b, D-1)

The electric power industry, the largest source of global-warming emissions in the United States (mainly through burning of coal to generate electricity) began a lobbying effort to shape future policy through the Edison Electric Institute, the industry's main lobbying group, and the Electric Power Research Institute, a major policy research arm of the industry. The Edison Institute during February 2007 dropped its long-standing opposition to legally required emissions limits. The power industry also was pressing for legislation that would allow surcharges on customers' electric bills to cover most of the costs of cleaning up power-plant emissions. The American Gas Association also came out in favor of modest, economy-wide regulations.

By 2007, ten U.S. states had started their own cap-and-trade program, the Regional Greenhouse Gas Initiative. Starting in 2009, each state will receive a set number of carbon credits for its power plants and will require each plant to either conserve energy or purchase enough allowances to cover its total emissions (Fairfield 2007).

On February 26, 2007, two days after former vice president Al Gore's *An Inconvenient Truth* won an Oscar for best documentary, the governors of California, Oregon, New Mexico, Arizona, and Washington jointly announced plans for a cap-and-trade program that allocates carbon credits to companies, for sale or trade. New England states and New York have adopted a similar plan (Fialka 2007a, A-8).

## EMISSION CAPS IN THE UNITED STATES: POLITICAL CONTEXT

Early in 2007, several economists and corporate leaders, as well as environmentalists, told two Senate committees that the United States should quickly impose mandatory restrictions on greenhouse-gas emissions and display leadership on the climate-change issue globally. "Uncontrolled climate change constitutes a risk that we as a global community cannot afford to take," former World Bank economist chief economist Sir Nicholas Stern told the Senate Energy and Natural Resources Committee (Pegg 2007a). Stern, head of the British Government Economic Service and lead author of the most comprehensive economic review of climate change to date, said climate change is a global problem that "requires a global response." "Equity demands that rich countries takes the lead," Stern said. "It is they who are responsible for the bulk of the problem. And it is the poor countries will be hit earliest and hardest" (Pegg 2007a).

*The Stern Review*, issued during October 2006, warned that unabated climate change will sharply impact societies and ecosystems across the planet and could cost the world 5 to 20 percent of gross domestic product (GDP) annually (Pegg 2007a). The costs of acting now, however, can be limited to about 1 percent of annual global GDP, according to the report. "That amount will not slow growth," Stern said. "The later we leave it, the greater the risks and the higher will be costs of controlling them" (Pegg 2007a).

Some lawmakers evoked concern about the United States enacting mandatory emission reductions without obligations from rapidly industrializing nations that are exempt from the Kyoto Protocol, most notably China and India. Sen. Pete Domenici (R-N.Mex.) said he is "more and more fearful" of the consequences of the United States acting alone. China uses more coal than the United States, he said, and is opening a new coal-fired power plant virtually every other week. "Somebody in a big leadership role has to get together with the Chinese and the Indians, and decide whether they have a stake or not—and if they do then we should try to do something together," Domenici told colleagues (Pegg 2007a).

Stern urged Congress to recognize that China is taking steps to tackle climate change. He said that China is no longer deforesting, has a plan to cut its energy intensity 20 percent in five years, and has imposed an export tax on steel and other energy intensive goods. "You cannot sell an American car in China because they don't meet emissions [standards]," Stern said. "It is not correct to say that China is doing nothing." Stern added that China, India, and other nations also are watching for the United States to act more aggressively and will go faster if the United States demonstrates leadership. "Leadership in the world's largest markets sets the pace elsewhere," Stern said (Pegg 2007a).

During his remarks before the energy committee, Stern said that climate change is "the biggest market failure the world has ever seen." He continued: "People should pay in the prices they face for the costs of their actions—in this case costs to the climate.... Pricing carbon directly through either tax or carbon trading or implicitly through regulation is fundamental to a policy response." Stern called for a mix of policies, including carbon taxes and emissions trading, but acknowledged that the latter is more politically feasible, even though it is a far more complex undertaking (Pegg 2007a).

## PERSONAL CARBON BUDGETS: CAP-AND-TRADE FOR INDIVIDUALS

Some activists have advocated a cap-and-trade market at the personal level. Each person would receive the same number of carbon credits and have the right to buy and sell them. Ride a bicycle to work, for example, and get paid by your neighbor who can't leave his Hummer behind. The proposed market would extend worldwide and would be designed to equalize anticipated effects of global warming between rich and poor nations (Walsh 2007a).

The infrastructure for cap-and-trade at the personal level is being created with a credit-card device that would allocate each individual a carbon allowance or ration, credits that could be traded on the Internet. The carbon credit card as presently envisioned would be charged for energy-related purchases (such as residential electricity, natural gas, or heating oil) and transportation. Still in the future is a card that might score the energy used to create and transport every purchase—those sports shoes made in

Vietnam, for example, or that winter watermelon shipped to New York City from Venezuela. Carbon allowances could be steadily reduced to foster energy conservation and reduce greenhouse-gas emissions at the personal level, as libertarians, no doubt, rail against them as another invasion of big government and e-hustlers sell bogus credits on a black market (Hillman, Fawcett, and Rajan 2007, 194–98).

Citizens in the Netherlands have been using a credit card that converts purchases into funding for climate restoration. The carbon-footprint credit card, issued by Rabobank, the largest commercial bank in the Netherlands, was being used by 1.1 million customers by 2007. Calculations convert the carbon dioxide emission value of purchases into a cash amount. The bank contributes to projects developed by the World Wildlife Fund that counter effects of greenhouse-gas emissions (Environment News Service 2007dd). "Taken this way, offsetting is cutting edge in two ways," said Barbera van der Hoek, who leads the Dutch WWF climate and energy program. "In the Netherlands it helps people to become aware of the climate change impact of their own buying behavior. In developing countries Gold Standard projects help build local sustainability and a positive change of the energy system" (Environment News Service 2007dd).

Toronto mayor David Miller described a hybrid environmental footprint calculator and web-based social network, Zerofootprint Toronto, which illustrates to users the impact every aspect of their daily lives has on the environment, then compares the results with others.

## CARBON OFFSETS: ARE THEY REAL?

By 2007, Americans were imitating Europeans by gobbling up carbon offsets, by which a person or an organization can pay for carbon-creating activities by donating to projects that offset them, such as planting trees, clean-energy projects, or pollution control. Supporters maintain that offsets raise money for good causes and demonstrate the necessity of restricting carbon-producing activities on a personal level. Critics claim that offsets are "green lite," allowing carbon dioxide–creating consumer activities (and their emissions) to continue, while discharging guilt without actually curtailing greenhouse-gas emissions.

The market for carbon offsets in the United States grew by 80 percent between 2005 and 2006, to $55 million annually, according to a report from New Carbon Finance and Ecosystem Marketplace, despite the fact that no standards exist to define precisely what these instruments buy other than freedom from climatic guilt. One utility, American Electric Power, agreed to offset emissions of about 4.6 million tons of carbon dioxide by paying for projects that reduce the amount of methane—a powerful pollutant—seeping from farm manure (Fahrenthold and Mufson 2007, A-1).

Al Gore, for example, has been roundly criticized for offsetting a house with eleven bathrooms and flying around the world on greenhouse-gas-spewing jets to show his film *An Inconvenient Truth*. The way to counter global warming, many argue, is to actually *reduce* greenhouse gases.

Perhaps Gore, if he means what he says, should give up his mansion for a much smaller house and show his film over the Internet.

"There needs to be more standardization, more verification and more assurances for the consumer that the offsets are real," conceded Ricardo Bayon, director of Ecosystem Marketplace. A number of organizations, including the Center for Resource Solutions in San Francisco and the Climate Group, based in Britain, raced to establish certification standards. "This voluntary stuff is an interim measure," said Judi Greenwald of the Pew Center on Global Climate Change. "But it is certainly better than doing nothing" (Kher 2007).

For those who are seriously rich and *really* feeling guilty about their gigantic carbon footprints, offsets are available for that private jet that uses fifteen times the fuel per passenger mile of a commercial flight and about forty-five times the amount of an average automobile with one passenger. Mega-mansions can be green-certified (solar-heated swimming pools included). With offsets, no one has to drive a small car, live in a two-room shack, or ride a bicycle. Jets.com, a private jet service, has joined with the Carbon Fund to bill private jet owners for their emissions; V1 Jets International has a similar program. Guilt can be discharged on the cheap: a round-trip flight between Fort Lauderdale, Florida, and Boston costs about $20,000—including a $74 carbon offset for the 13 metric tons of $CO_2$ produced (Frank 2007, W-2).

## ROCK AND ROLL AND CARBON OFFSETS

Offsets have become popular with carbon-conscious rock-'n'-roll musicians seeking to counter criticism of their worldwide tours hauling massive stage props and staffs of roadies. The rock band Pearl Jam announced during July 2006 that it would offset greenhouse gases released during its 2006 world tour (sixty-nine stops, from April 20 in London, England, through November 25 in Perth, Australia) from the trucks, buses, airplane travel, hotel rooms, concert venues, and fans driving to and from their concerts—by providing funding to nine nonprofit organizations that help to reduce global warming (Environment News Service 2006c).

The music-industry trade journal *Billboard* published a lengthy front-page article detailing how musical groups had joined with Future Forests to plant trees to make up for the ecological costs of their activities. The *Billboard* article said:

> Numerous artists and music companies are taking a leading role in an environmental program that aims to combat global warming. Foo Fighters, Coldplay, Gorillaz, Kylie Minogue, Shaggy, Mis-teeq, Dido, Neneh Cherry, and Sting—to name a few acts—have linked with Future Forests, a London-based, for-profit company, to ensure that their activities do not exacerbate the ecological problems facing the planet. (Masson 2003, 1)

"There are serious problems storing up for us now," Future Forests founder and chairman Dan Morrell was quoted as saying. "But basically, by planting trees, we can make everything we do carbon-neutral, and that's at least a start in fighting those problems" (Masson 2003, 1).

The Rolling Stones performed a special free concert to raise awareness about global warming on February 6, 2003, at the Los Angeles Staples Center. The Natural Resources Defense Council (NRDC) joined with the Rolling Stones to stage the event, hoping, it said, to turn up the heat on the Bush administration's inaction regarding global warming. The partners also emphasized opportunities to start addressing the problem. A private anonymous donor absorbed the costs of the concert; tickets were awarded in a lottery-style sweepstakes.

The Live Earth concerts on July 7, 2007, on all seven continents (Antarctica included) brought together more than a hundred music acts to educate people about global warming. The twenty-four-hour series of concerts began in Sydney and continued in Tokyo, Shanghai, Istanbul, Johannesburg, London, Hamburg, Rio de Janeiro, Washington, D.C., and Antarctica before concluding in New York City (Environment News Service 2007r).

The concert in Washington was added at the last minute under the sponsorship of the National Museum of the American Indian, two blocks from the U.S. Capitol, invoking "Mother Earth." Vice President Gore had tried to organize a concert on the National Mall, but was blocked by a small group of Senate Republicans led by Senator Inhofe, who insists

**Figure 8.3. Live Earth concert at Wembley Stadium, London, July 7, 2007.**
Source: Wikipedia Commons.

that global warming is a "hoax." "There were some naysayers who tried to say, 'No, you cannot have a concert on the Mall,'" Gore said. "But the cavalry didn't ride to the rescue, the American Indians did." The concert, held with the Capitol building as a backdrop, featured a rare concert by country superstar Garth Brooks, one of very few following his retirement in 2001, alongside his wife Trisha Yearwood, also a country star.

Live Earth's critics contended that the concerts had no achievable agenda and used copious amounts of fossil fuels to assemble jet-setting rock stars in fuel-guzzling airliners, using huge amounts of electricity to produce little more than a lot of noise. But organizers kept a tally of energy use and used proceeds (tickets cost as much as $350 each) to distribute power-efficient light bulbs and otherwise offset the shows' greenhouse-gas emissions. More than 150 artists performed at the eight concerts (Associated Press 2007e).

John Buckley of Carbon Footprint, an organization that helps companies reduce their carbon dioxide emissions, said that Live Earth produced about 74,500 tons of carbon dioxide. "We would have to plant 100,000 trees to offset the effect of Live Earth," he said (Cooper 2007, D-4). London newspapers criticized the energy expenditures of many performers, some of whom flew private planes halfway around the world to play in the concerts. "The Artists Formerly Known as Huge Carbon Footprints" headlined London's *Guardian*. The *Daily Mail* noted that Madonna, one of the headliners in London, has several gas-guzzlers in her garage: "Waiting in the garage at home, she has a Mercedes Maybach, two Range Rovers, an Audi A8 and a Mini Cooper S" (Riding 2007).

Gore replied that the concerts created "a critical mass of opinion worldwide [that] will push the world across a tipping point beyond which political and business and civic leaders across the spectrum will begin offering genuinely meaningful solutions to the climate crisis. I think that's a realistic hope, and it's greatly needed" (Sisario 2007).

The same day as the concerts, a heat wave scorched the western United States. Missoula, Montana, experienced its highest temperature on record (107°F) and a wave of forest fires, aided by record heat and drought, spread over the region. Yellowstone National Park and Montana state fisheries managers banned fishing on some rivers between noon and 6 P.M. because water temperatures above 73°F could stress and even kill fish.

## CARBON NEUTRALITY: "ECO-LITE"?

Along with purchase of carbon offsets came pledges of "carbon neutrality"—some from political campaigns, another carbon-intensive activity that hauls large numbers of people long distances, often in jet aircraft. By 2007, several politicians, including John Edwards and Hillary Clinton, were running so-called carbon-neutral campaigns, at least part of the time. Carbon neutrality usually was achieved through attention to conservation and purchase of offsets.

A lengthening list of big businesses, including several large banks, London's taxi fleet, and even a few airlines also asserted carbon neutrality. For example, Silverjet, a luxury trans-Atlantic air carrier, promoted itself as the first fully carbon-neutral airline because it donated about $28 from each round-trip ticket to a fund for projects as esoteric as fertilizing the oceans with iron so that algae can pull more carbon dioxide out of the atmosphere. In theory, said representatives of Silverjet, such projects may eventually eliminate as much carbon dioxide as the airline generates, or about 1.2 tons per passenger per trip. The "cure" was on future speculation, however, while greenhouse gases still were being emitted in the present tense. An entire consulting industry grew to calculate offsets and bill the companies, handling the transactions with a fee attached. By early 2007, *Business Week* magazine said that the trade in offsets was more than $100 million a year "and growing blazingly fast" (Revkin 2007). In the meantime, worldwide greenhouse-gas levels continued to rise.

"The worst of the carbon-offset programs resemble the Catholic Church's sale of indulgences back before the Reformation," asserted Denis Hayes, president of the Bullitt Foundation, which specializes in environmental projects. "Instead of reducing their carbon footprints, people take private jets and stretch limos, and then think they can buy an indulgence to forgive their sins." Hayes added, "This whole game is badly in need of a modern Martin Luther" (Revkin 2007).

Michael R. Solomon, author of *Consumer Behavior: Buying, Having and Being* and a professor at Auburn University, said he was not surprised by the appeal of ersatz carbon neutrality. "Consumers are always going to gravitate toward a more parsimonious solution that requires less behavioral change," he said. "We know that new products or ideas are more likely to be adopted if they don't require us to alter our routines very much." He sees danger ahead "if we become trained to substitute dollars for deeds— kind of an 'I gave at the office' prescription for the environment" (Revkin 2007).

Charles Komanoff, an energy economist in New York City, said the commercial market in climate neutrality could harm effects to actually reduce carbon emissions. By suggesting an easy way out, assertions of carbon neutrality might blunt public support for necessary efforts in the long run, including a mandatory limit on emissions or a tax on the fuels that generate greenhouse gases. "There isn't a single American household above the poverty line that couldn't cut their $CO_2$ at least 25 percent in six months through a straightforward series of fairly simple and terrifically cost-effective measures," he said (Revkin 2007).

"There is a very common mind-set right now which holds that all that we're going to need to do to avert the large-scale planetary catastrophes upon us is make slightly different shopping decisions," said Alex Steffen, the executive editor of Worldchanging.com, a website devoted to sustainability issues (Williams 2007). Steffen said that the only valid way to deal with global warming is to actually *reduce* emissions in real time. Instead of jetting off to a second home after paying offsets, own one home and stay there, Steffen advised.

Paul Hawken, author and environmental activist, criticized the false promise of chic green consumerism, which he called oxymoronic. "We turn toward the consumption part because that's where the money is," Hawken said. "We tend not to look at the 'less' part. So you get these anomalies like 10,000-square-foot 'green' homes being built by a hedge fund manager in Aspen. Or 'green' fashion shows. Fashion is the deliberate inculcation of obsolescence." He continued: "The fruit at Whole Foods in winter, flown in from Chile on a 747—it's a complete joke. The idea that we should have raspberries in January, it doesn't matter if they're organic. It's diabolically stupid" (Williams 2007).

As the debate continued over the idea's usefulness in the real climatic world, the House of Representatives during 2007 made plans to go carbon neutral. House chief administrative officer Dan Beard, who was directed in March 2007 by Speaker Nancy Pelosi to find ways to make the House side of the Capitol carbon neutral, has worked to omit coal from the fuel mix that heats and cools the Capitol and nearby buildings. In addition, Beard's office also changed twelve thousand desk lamps in House office buildings to compact fluorescent bulbs, then installed dimmers. Next, Beard said, the House will to buy its electricity from renewable sources such as wind and solar through an arrangement with Pepco, a Washington, D.C., electric utility. After the House replaces coal with natural gas at the power plant, it will reduce its annual carbon dioxide emissions by 75 percent. To deal with the remaining 25 percent and become carbon neutral by 2010, the House may buy offset credits or invest in conservation projects (Layton 2007, A-17).

The Capitol power plant, four blocks from House office buildings, has burned coal since it opened in 1910 and is the only remaining coal-burning facility in the District. The power plant (which produces no electricity) consumes coal as half its fuel (the rest is oil and natural gas) to generate steam and chilled water to heat and cool the Capitol, the Supreme Court, the Library of Congress, and nineteen other structures (Layton 2007, A-17).

## THE AGENDA IN THE U.S. CONGRESS

By late 2007, the House of Representatives and the Senate were sketching the outline of legislative battles to come. On October 16, 2007, the House passed an unenforceable "sense of Congress" resolution setting a goal to expand renewable energy production in the United States to at least 25 percent of energy use by 2025. At the time, renewable energy (including hydroelectric) supplied about 6 percent (Environment News Service 2007s).

A bill introduced a week later in the Senate would require mandatory limits on greenhouse gases on the order of more than 60 percent by 2050. The proposal, introduced by Connecticut Independent Joe Lieberman and Virginia Republican John Warner, would impose greenhouse-gas limits on three-quarters of the economy, limiting emissions from electric power, transportation, and manufacturing industries to 1990 levels by 2020. The America's Climate Security Act was introduced into the Senate Environment and Public Works Committee's global warming subcommittee.

"The basic difference between the administration approach and our approach is that we feel that voluntary [measures] will not achieve the goals—the leadership—that the United States of America must simply take on this issue," said Warner (Pegg 2007b). The proposal was given some praise by environmental groups, such as Environmental Defense, the NRDC, and the National Wildlife Federation. "Lieberman and Warner have paved the way for a historic committee vote on a bill that promises to make great strides toward climate security and economic growth," said Steve Cochran, the national climate campaign director for Environmental Defense. "Thanks to their thoughtful approach we're moving beyond talk and quickly toward action" (Pegg 2007b).

Other groups, including Friends of the Earth and the Sierra Club, criticized the bill. The legislation "falls far short of our responsibility as one of the world's largest global warming emitters," said Friends of the Earth president Brent Blackwelder. The plan misses "critical targets and timetables," he said. According to the Environment News Service, Blackwelder "criticized the free allocation of some of the emission permits as well as the failure to help the world's poor adapt to climate change" (Pegg 2007b).

## AN ECOWARRIOR IN THE WHITE HOUSE?

By 2007, the United States was walking away from George W. Bush's denial of global warming. Perhaps, in a tepid way, Bush even was walking away from himself. The President, who once had off-handedly called global warming a fantasy not supported by "sound science," called upon the fifteen largest greenhouse-gas-emitting nations to set limits. "The United States takes this issue seriously," Bush said. This was the same man who, a couple of years earlier, took Michael Crichton's novel *State of Fear* as his standard on the issue (Associated Press 2007b). Many environmentalists regarded Bush's new stance as eco-deception, however, because of its voluntary nature. Friends of the Earth president Blackwelder called the proposal "a complete charade. It is an attempt to make the Bush administration look like it takes global warming seriously without actually doing anything to curb emissions" (Associated Press 2007b).

Support was growing worldwide for a target allowing global temperatures to increase no more than 2°C (3.6°F). This would require a global reduction in emissions of at least 80 percent below 1990 levels by 2050. The Bush administration flatly rejected that proposal, instead supporting goals that were "aspirational" (Associated Press 2007b). In real English, that meant voluntary, with no concrete targets or dates, no enforcement, and no penalties for noncompliance. Bush also rejected a global cap-and-trade carbon-trading program. The White House further expressed opposition to the energy-efficiency targets advocated by the European Union.

The next day, recalling Bush's Iraq-war rhetoric, environmentalists joked about whether global warming would be aspired over "to stay the course" or "until the job is done." If voluntary action was so effective, some suggested trying it with war funding. Put a tin cup outside the door at 1600 Pennsylvania Avenue and see what you get by the end of the day.

## FORMATION OF A U.S. CARBON REGISTRY

By mid-2007, as the number of companies supporting mandatory greenhouse-gas limits grew, thirty-one states, representing more than 70 percent of U.S. population, announced formation of the Climate Registry, the first effort to systematically measure greenhouse gases in the United States. The list of founding states included Arizona, California, Colorado, Connecticut, Delaware, Florida, Hawaii, Illinois, Kansas, Maine, Maryland, Massachusetts, Michigan, Minnesota, Missouri, Montana, New Hampshire, New Jersey, New Mexico, New York, North Carolina, Ohio, Oregon, Pennsylvania, Rhode Island, South Carolina, Utah, Vermont, Washington, Wisconsin, and Wyoming. Two Canadian provinces, British Columbia and Manitoba, also joined. Heretofore, greenhouse-gas emissions were not required to be measured or reported in the United States.

"States cannot wait any longer for leadership on global warming from the federal government," said Illinois governor Rod Blagojevich. "States are creating a system that gives businesses and organizations an opportunity to step up to the plate and take responsibility for reducing their greenhouse gas emissions" (Environment News Service 2007y).

The Climate Registry was established to measure greenhouse-gas emissions across borders and industries, a first step toward developing programs to reduce them. "You have to be able to count carbon pollution in order to cut [it]," said Frances Beinecke, president of the NRDC. "The Registry gives business and policymakers an essential accounting tool for tracking the success of the many emerging global warming emission reduction initiatives that are blossoming across the country" (Environment News Service 2007y).

Florida, with most of its real-estate assets on its coastlines, is looking down the barrel of a climate-change cannon, as seas slowly rise and the land subsides, accelerated, at times, by severe hurricanes. The cost of inaction there would be especially high. That state's Republican governor, Charlie Crist, on July 13, 2007, signed executive orders that committed that state, with the fourth-largest population in the United States, to match California's standards for greenhouse-gas reductions in emissions from automobiles. Crist said he would also implement a policy to reduce Florida's greenhouse-gas emissions to 20 percent of 1990 levels by 2050 (an 80 percent reduction), the amount that could contain global warming in the long range.

Governor Crist also pledged legal action to raise energy-conservation standards in new buildings, while requiring utilities to acquire 20 percent of their electricity production from renewable sources. Asked whether he was worried about the cost of meeting those targets, he cited the effect on the state's tourism industry, saying "there's an economic cost to not doing these things." "It is a very impressive package of actions from a 'red' state and from a Republican governor taking a leadership position on global warming," said David Doniger, policy director for climate at the NRDC (Mufson 2007b, D-1).

## ECO-CONVERSION OF SPECIAL INTERESTS?

As the political center of gravity changed in the United States vis-à-vis solutions to global warming, we may be addressing a problem that James E. Hansen, director of NASA's Goddard Institute for Space Studies, has raised in many of his public presentations: the resistance of special interests. At long last, the idea that we share the same Earth was becoming the gold standard of political discourse on the issue. By 2007, some of the special interests (even a few oil companies) were realizing that without a sustainable habitat, we do no business and we tell no stories. Hansen had long warned:

> The greatest obstacle to solving the climate crisis is the "special interests." As long as the coffers of our elected representatives can be filled by special interests, the latter will keep calling the tunes. Until there is true campaign finance reform, the special interests will continue to make a mockery of the central proposition of our democracy, that the commonest of men should have a vote equal in weight to that of the richest, most powerful citizen. (Hansen, personal communication, March 8, 2007)

By mid-2007, a *New York Times*–CBS News poll found that 78 percent of Americans believed that global warming required action "right away"— and none too soon. Hansen's long-term forecast: ten more years of "business as usual" greenhouse-gas emissions and we'll very probably cross a "tipping point." At that time, humanity will lose any chance of turning back accelerating feedbacks that lead to runaway heating of Earth's atmosphere. Mark your calendars; Hansen has been forecasting global greenhouse weather for more than twenty-five years, and his record is very good, even as he has had to run rear-guard actions to keep three Republican U.S. Presidents backed by fossil-fuel partisans from de-funding his laboratory and censoring his statements.

## CORPORATE SUSTAINABILITY OFFICERS

Some companies are appointing "chief sustainability officers" (CSOs). These officers, who may go under a title like "vice president for environmental affairs," join with vendors and customers to create and market green products. Dow Chemical's first CSO, David E. Kepler, has been meeting with Dow's technology, manufacturing, and finance leaders about alternative fuels and green products. "We usually agree," Kepler said. "But if a critical environmental issue is in dispute, I'll prevail" (Deutsch 2007a).

Linda J. Fisher, the CSO at DuPont, weighed in against purchase of a company that was not in a "sustainable" business. "We're building sustainability into the acquisition criteria," she said. When two business chiefs at General Electric opposed the cost of developing environmentally friendly products, Jeffrey R. Immelt, GE's chairman, gave Lorraine Bol-singer, vice president of GE's Ecomagination business, the research money. "I have an open door to get projects funded," she said (Deutsch 2007a).

Owens-Corning named Frank O'Brien-Bernini, its research chief since 2001, to the post of chief research and development and sustainability offi-cer. He now uses what he calls the "lens of sustainability" to prioritize research. He helped husband development of a machine to ease insulation of attics. Owens-Corning developed it after research showed that drafty attics are prime culprits in greenhouse-gas emissions but that the "hassle factor" kept homeowners from addressing the situation. "I drive innova-tion around products and processes," O'Brien-Bernini said. "And I make sure that our claims are backed by deep, deep science" (Deutsch 2007a). Elizabeth A. Lowery, vice president for environment, energy, and safety policy at General Motors, helped to brief reporters, shareholders, and envi-ronmentalists on the Chevy Volt, an electric car.

Stephen Lane, who jokes that he is the "Al Gore of Citigroup," has a full-time job coaxing energy savings out of the 340,000-employee, world-wide financial-services company. Lane oversaw an inventory of energy use in all of the company's facilities. "What you can't measure, you can't man-age," he says (Carlton 2007, B-1), He also set company policies that gov-ern everything from installing solar energy and timed lighting in bank branches to convincing employees to shut off unused lights and climb stairs instead of using escalators (which are now stopped during nonbusi-ness hours). Other banks are taking similar measures. The Hong Kong and Shanghai Banking Corporation, for example, has opened a "green" prototype branch in Greece, New York, as a model for the company's four hundred branches in the United States. Lane oversees a $10 billion Cit-igroup plan to reduce its carbon footprint to 10 percent below 2005 levels by 2011 (Carlton 2007, B-8).

By the end of 2007, more than 2,400 companies in the United States were reporting their carbon emissions and energy costs through the Lon-don-based Carbon Disclosure Project, a nonprofit group of 315 corpora-tions and institutional investors. Some are household names, such as Coca-Cola and Wal-Mart (which was among the first to join). Wal-Mart during 2007 began requiring its suppliers to report and reduce their carbon foot-prints. Wal-Mart initiated its program by asking Oakhurst Dairy of Portland, Maine, to measure the carbon footprint of a case of milk. Dell, the computer maker, announced in 2008 that it will begin to neutralize the carbon impact of its operations around the world (Cox News Service 2007, D-1). By the end of 2007, Procter & Gamble, Unilever, Tesco, and Nestlé were requiring their suppliers to disclose carbon dioxide emissions and global-warming miti-gation strategies. The companies are among several that have formed the Supply Chain Leadership Coalition, which cooperates with the Carbon Dis-closure Project. Eventually, products may be labeled with carbon-emission information. In 2007, Cadbury Schweppes was making plans to print such information on its chocolate bars (Spencer 2007, A-7).

The Frito-Lay plant in Casa Grande, Arizona, has been implementing plans to take itself nearly completely off the power grid by 2010, using several strategies. One is new filters that recycle most of the water used to wash and rinse potatoes and corn destined for snack chips. The remaining sludge will then be used as a source of methane to power the factory's boiler. The factory also plans to install 50 acres of solar concentrators to be used for power, along with a biomass generator burning agricultural waste. Plans are to reduce electricity and water consumption by 90 percent, natural gas usage by 80 percent, and greenhouse-gas emissions 50 to 75 percent. These strategies, once adapted at the giant Casa Grande plant (the size of two football fields), will be considered at Frito Lay's thirty-seven other processing plants in the United States and Canada (Martin 2007, A-22).

## CONTINUED FEDERAL SUBSIDIES FOR COAL-FIRED POWER

Even as a growing political consensus formed to address rising levels of greenhouse gases, a Depression-era program designed to bring electricity to rural areas continued to use tax money to provide billions of dollars in low-interest loans to build coal-powered electrical power plants. The program was initiated during the mid-1930s by President Franklin D. Roosevelt to bring electricity to farms. Although this goal was achieved decades ago, rural electric cooperatives are still receiving federal subsidies as part of a $35 billion, ten-year program to build coal-fired power plants. These will spew into the atmosphere enough carbon dioxide to offset all other state and federal efforts to reduce U.S. greenhouse-gas emissions (Mufson 2007a, A-1).

When the federal Office of Management and Budget sought to end loans for new coal-fired power plants and limit loans for transmission projects, the National Rural Electric Cooperative Association deployed three thousand members to Capitol Hill during May 2007 to lobby Congress in favor of the program. The lobbyists asserted that the program was essential to the reliability of rural electricity (Mufson 2007a, A-1). In the meantime, environmentalists said that the program

> removes any pressure for the rural co-ops to promote energy efficiency or aggressively tap renewable resources. Rural co-ops rely on coal for 80 percent of their electricity, compared with 50 percent for the rest of the country, and electricity demand at rural co-ops is growing at twice the national rate. (Mufson 2007a, A-1)

In March 2008, the federal government suspended its loan program for new coal plants in rural communities through 2009 due in part to uncertainty over a suit filed by Earthjustice during July 2007. The suit cited the government's failure to consider global warming as it financed new coal plants.

## CHANGING THE TAX SYSTEM TO PENALIZE CARBON PRODUCTION

National taxation systems are being changed to discourage production of greenhouse gases, a trend that will ultimately reach the United States.

Many Europeans believe that the United States should sign both the Kyoto Protocol and a future agreement that will take its place in 2012. Many are suggesting a carbon tariff on exports to the United States from the European Union, the largest export market for U.S. goods, to force compliance. "A carbon tax is inevitable," said Jacques Chirac, president of France until 2007. "If it is European, and I believe it will be European, then it will all the same have a certain influence because it means that all the countries that do not accept the minimum obligations will be obliged to pay" (Bennhold 2007a).

A carbon tax need not increase the total tax burden. As in Sweden and other countries, taxes aimed at reducing carbon dioxide emissions could be used to replace income taxes, on a socially useful activity (labor), substituting a tax on activity that society should discourage (producing carbon dioxide). Gilbert Metcalf, a professor of economics at Tufts University,

> has shown how revenue from a carbon tax could be used to reduce payroll taxes in a way that would leave the distribution of total tax burden approximately unchanged. He proposes a tax of $15 per metric ton of carbon dioxide, together with a rebate of the federal payroll tax on the first $3,660 of earnings for each worker. (Mankiw 2007)

NASA's James Hansen has suggested "a gradually ... increasing price on carbon emissions" to drive energy efficiency improvements and promote innovative technologies (Hansen 2007b). He believes that a price on carbon emissions will help spur high-technology, high-paying employment in alternative fuels that will increase exports and improve the United States' balance of payments, hand-in-hand with a greater degree of energy independence that complements national security.

Hansen favors coupling taxation with reduction in greenhouse-gas emissions, a strategy that has become popular in Europe. He anticipates that technological innovation will increase energy efficiency in vehicles, buildings, appliances, and lighting that will accelerate as penalties for producing carbon dioxide increase. The carbon tax will guide investment away from development of "new" fossil fuels, such as tar sands and oil shales. Greenhouse-gas emissions by vehicles, buildings, and industrial processes should be addressed using efficiency standards as well as a carbon price, Hansen believes. "The carbon price must be ratcheted upward so as to optimize sustainable economic growth. Barriers to energy efficiency must be removed" (Hansen 2007b).

Hansen also believes that China, India, and other densely populated, rapidly developing countries that are exempt from the Kyoto Protocol must be factored into any workable worldwide plan to reduce greenhouse gases. Many of these countries will suffer some of the most intense effects of global warming, especially in large coastal cities, where rising sea levels will provide a strong incentive for moving promptly to technologies that do not release carbon dioxide (Hansen 2007b). Hansen has reminded legislators that while greenhouse-gas emission are increasing rapidly from China (which passed the United States as the single largest contributor per annum in 2007), the atmospheric burden is cumulative, so "the responsibility of the United States is more than three times that of any other single nation, and it will continue to be the largest for at least several decades" (Hansen 2007b; Environment News Service 2007i).

## SWEDEN AND DENMARK'S ENERGY TAXES

Sweden was the first country in the world to adopt a carbon tax, in 1991. Finland and the Netherlands soon followed. By 2007, nearly half of the Swedish income tax burden had been replaced by levies based in some way on fossil-fuel consumption. Sweden also levies an energy tax against all fossil fuels, and other taxes are used as incentives to control pollution, including a nitrogen oxides charge, a sulfur tax, and a tax on nuclear energy production. The Swedish taxation system unabashedly aims to control how people use energy. Automobiles in Sweden are taxed based on their greenhouse-gas emissions.

The Swedish energy tax varies widely among different products and types of fossil fuels, with by far the highest rates on gasoline. Responding to the new tax system, the use of biofuels (mainly wood-based) for "district" heating soared about 350 percent during the 1990s. Prices of wood-based fuels also fell dramatically as technology improved. Sweden's carbon tax has been credited with reducing fossil-fuel emissions about 20 percent.

Efficient industries are rewarded by Denmark's tax system, selling tax credits to those which are less efficient. Denmark's energy consumption has remained stable for thirty years (since the oil embargos of the mid-1970s) as its GDP has doubled, according to the Paris-based International Energy Agency (during the same period, U.S. energy use has risen 40 percent). The average Dane used 6,000 kilowatt-hours (kwh) of electricity in 2007, half the U.S. average (13,300 kwh) (Abboud 2007, A-13). Much of Denmark's energy sector is owned by nonprofit cooperatives with residents as shareholders. Denmark's social-welfare state values health care, free schools, and pensions over profits, free individual choice, and low taxes. Danish taxes have been raised to the point that they represent about half an average household's energy bills (Abboud 2007, A-13).

## CARBON TAX PROPOSALS IN THE UNITED STATES

As early as the 1980s, Colorado senator Timothy Wirth introduced legislation in the Senate Committee on Energy and Natural Resources (as the National Energy Policy Act of 1988) to place mandatory controls on industries that produce greenhouse gases. The proposal was received by lobbyists (the centurions of the special-interest state) like a proverbial lead balloon, as Senator Wirth's measures languished in committee. Wirth also sought to legislate stringent regulation of energy efficiency, especially in automobiles, which produce 40 percent of human-generated carbon dioxide in the United States. He favored limits on population growth and increased governmental support for alternative energies, such as wind and solar, that do not burn fossil fuels.

The U.S. tax code has been suggested as a tool to reduce fossil-fuel emissions, with such proposals as tax credits for energy-efficient appliances, changes in building codes focused on energy efficiency, energy efficiency standards for universities and health centers that receive federal grants, and tax incentives for fuel-efficient cars. The American Enterprise Institute, a conservative think tank, estimates that a tax of $15 per ton of carbon dioxide emitted would add about 14 cents to a gallon of gasoline and $1.63 per kilowatt-hour to the cost of electricity (Solomon 2007, A-4).

Carbon taxes already have been adopted in parts of the United States. Voters in Boulder, Colorado, home of the state's largest university, late in November 2006 approved the United States' first carbon tax, based on electricity usage. The tax, which took effect April 1, 2007, adds $16 a year to an average homeowner's electricity bill and $46 for businesses. It is collected by Boulder's main gas and electric utility, Xcel Energy, as the agent for Boulder's Office of Environmental Affairs. The tax revenue will help consumers "increase energy efficiency in homes and buildings, [aid a] switch to renewable energy, and reduce vehicle miles traveled," said Jonathan Koehn, the city's environmental affairs manager. The Boulder environmental sustainability coordinator, Sarah Van Pelt, said residents who used alternative sources of electricity like wind power would receive a discount on the tax based on the amount of the alternative power used (Kelley 2006). A total of 5,600 residents and 210 businesses used wind power in 2006, she said.

The state of Oregon in 2001 began to assess a 3 percent fee on electricity bills issued by its two largest private utilities. Revenue from this tax is transferred to the Energy Trust of Oregon, a nonprofit organization, rather than the state government. The trust distributes cash incentives to businesses and residents for using alternative sources, including solar and wind power and biomass energy, as well as for structural improvements to improve energy efficiency (Kelley 2006).

## CARBON TAX COSTS AND BENEFITS

Chris Flaven of the Worldwatch Institute has proposed a worldwide carbon tax of $50 a ton. He also proposes paying 10 percent of the tax into a fund to subsidize development of new technologies that will reduce emissions of greenhouse gases. A study of greenhouse-gas emission reduction strategies in Chicago supported a "$CO_2$ Fund ... which would pool resources from the state and private industry and make funds available in low-income communities in order to encourage emissions reductions" (Illinois Environmental Protection Agency n.d.) In *Earth in the Balance* (1992), Al Gore supported a fossil fuel tax, with proceeds to be paid into an Environmental Security Trust Fund, "which would be used to subsidize the purchase by consumers of environmentally benign technologies, such as low-energy light bulbs or high-mileage automobiles" (Gore 1992, 349).

William D. Nordhaus, one of the first to estimate the costs and benefits of global warming, has modeled the economic effects of carbon taxes ranging from $5 to almost $450 per ton (Nordhaus 1991, 56). He provided a range of carbon-tax rates with expected reductions in fossil fuel use and costs to GDP in the United States and calculated that a carbon tax of $13 a ton would reduce carbon emissions 6 percent (while reducing greenhouse gases 10 percent) at a negligible cost to economic output. Nordhaus expects a carbon tax of $98 a ton to reduce greenhouse gases 40 percent, with a half a percentage point decrease in production. His highest tax estimate, $448 per ton, would reduce greenhouse gases 90 percent and reduce GDP by more than 4 percent (Cline 1992, 168).

Nordhaus estimated that a $5-per-ton tax would raise the price of coal 10 percent, the price of oil 2.8 percent, and the price of gasoline 1.2 percent. Applied worldwide, the same level of carbon taxation would, according to Nordhaus's models, reduce greenhouse-gas emissions 10 percent, provide $10 billion in tax revenue, and add $4 billion per year to the global economy.

The numbers calculated by Nordhaus indicate that achieving the kind of greenhouse-gas reductions sought by the Intergovernmental Panel on Climate Change (IPCC) would impose a high cost in economic dislocation, especially if the changes are undertaken on a crash basis. A $100-per-ton tax on carbon emissions, for example, would raise the price of coal about 200 percent, the price of oil about 50 percent, and the price of gasoline between 23 and 24 percent. The $100 tax would reduce greenhouse-gas emissions by an estimated 40 to 45 percent, would provide $125 billion in tax revenues, and would decrease global "net benefits" $114 billion, according to Nordhaus.

Nordhaus's estimates are but one set of many, which vary widely. Michael E. Mann and Richard J. Richels, to cite but one example, contend that a carbon tax of $250 a ton would be necessary to suppress carbon dioxide emissions by 20 percent. Such a tax would add about 75 cents to the cost of a gallon of gasoline and $30 to the cost of a barrel of oil (Cline 1992, 147).

More than two thousand economists, including six Nobel laureates, endorsed a statement on global warming during 1997 that called for the use of market mechanisms, including carbon taxes, to move world economies away from fossil fuels:

> As economists we believe that global climate change carries with it significant environmental, economic, social, and geopolitical risks, and that preventive steps are justified. ... Economic studies have found that there are many potential policies to reduce greenhouse-gas emissions for which the total benefits outweigh the total costs. For the United States in particular, sound economic analysis shows that there are policy options that would slow climate change without harming American living standards and these measures may in fact improve U.S productivity in the longer run. (Economists' Letter 1997)

## CITIES ORGANIZE AGAINST GLOBAL WARMING

During May 2007, at the "C40 Large Cities Climate Summit," which met in New York City, a group of the world's largest cities committed to addressing climate change. Mayors from across the United States and around the world attended, including those from Bangkok, Berlin, Bogota, Chicago, Copenhagen, Delhi, Houston, Istanbul, Johannesburg, Mexico City, Rio de Janeiro, Rome, São Paulo, Seoul, Sydney, Tokyo, Toronto, and Vancouver. The mayors, their senior staff members, and business leaders shared projects showing how they were reducing greenhouse-gas emissions and conserving energy.

In Los Angeles, Mayor Antonio Villaraigosa, in partnership with the Los Angeles City Council and environmental leaders, unveiled "Green L.A.—An Action Plan to Lead the Nation in Fighting Global Warming." Villaraigosa pledged to reduce his city's carbon footprint by 35 percent below 1990 levels by 2035, the most ambitious goal yet set by a major American city. Los Angeles also plans to increase its use of renewable energy to 35 percent by 2020, much of it through changes at its municipal electrical utility, the largest in the country (Environment News Service 2007c).

By early 2008, 780 cities in the United States from Boston to Portland, Oregon, with a combined population of more than 75 million, had pledged to meet Kyoto Protocol standards, as local officials sped ahead of federal policy on global warming. A nationwide poll released at that time indicated that a third of Americans regarded global warming the world's single most pressing environmental problem, double the number from a year earlier, according to a poll conducted by the *Washington Post*, ABC News, and Stanford University. Greg Nickels, Seattle's mayor, started the U.S. Mayors Climate Protection Agreement in 2005. In the next year, U.S. fossil-fuel-related emissions fell 1.3 percent to 5.88 billion metric tons.

At the Seattle Climate Protection Summit during November 2007, more than a hundred U.S. mayors called for a federal partnership against energy dependence and global warming. "We are showing what is possible in light of climate change at the local level, but to reach our goal of 80 percent reductions in greenhouse gases by 2050, we need strong support from the federal government," said Nickels (Environment News Service 2007aa). New York mayor Michael Bloomberg said, "Climate change presents a national security imperative for us, because our dependence on foreign oil has entangled our interests with tyrants and increased our exposure to terrorism. It's also an economic imperative, because clean energy is going to be the oil gusher of the 21st century" (Environment News Service 2007aa). Bloomberg called for a pollution fee (a carbon tax) to discourage practices that generate greenhouse gases. "As long as greenhouse gas pollution is free, it will be abundant," he said. "If we want to reduce it, there has to be a cost for producing it. The voluntary targets suggested by President Bush would be like voluntary speed limits—doomed to fail" (Environment News Service 2007aa).

In Austin, Texas, energy efficiency standards were raised for homes, requiring a 60 percent reduction in energy use by 2015. Chicago is trying waterless urinals and has planted thousands of trees. Philadelphia has been replacing black tar paper roofs atop old row houses with snow-white, high-reflection composites. Keene, New Hampshire, requires parents waiting for children at schools to turn off their car engines. In Portland,

Oregon, carbon emissions have been reduced to 1990 levels, and water flowing through its drinking-water system also generates hydroelectricity (Faiola and Shulman 2007, A-1). Mayors of at least 134 U.S. cities by 2007 were (or soon would be) using more energy-efficient lighting in public buildings, streetlights, parks, traffic signals, and other places. Many city governments' auto fleets had converted to alternative fuels or hybrid-electric technology. Chicago's "Smart Bulb Program" by 2007 had distributed half a million free compact fluorescent light bulbs to residents.

Using the AlbuquerqueGreen program, that city reduced municipal natural-gas usage 42 percent and cut greenhouse-gas emissions 67 percent between 2000 and 2006. AlbuquerqueGreen promotes growth of green-tech companies, bicycle use, and pedestrian-friendly streets. Albuquerque requires that all new buildings be designed to be carbon neutral, with architecture suitable for 100 percent renewable energy use by 2030 (Environment News Service 2007bb).

## THINKING GLOBALLY AND ACTING LOCALLY IN NEW HAMPSHIRE

New Hampshire town residents during the spring of 2007 considered a state referendum on climate change. Of the 234 incorporated cities and towns in New Hampshire, 180 voted on a resolution petitioning the federal government to deal with climate change, including research that will create "innovative energy technologies" (Zezima 2007). As of March 18, 2007, 134 towns had passed the initiative at their town meetings. The resolution is nonbinding, but organizers hoped it would induce presidential candidates to raise the climate-change issue during the New Hampshire presidential primary in 2008. "We're trying to bring to the attention of presidential candidates that we are concerned about this in little purple New Hampshire," said Don Martin, 61, a real estate agent in Bristol who helped collect signatures to put the initiative on the agenda in his town, where it passed by a wide margin. "New Hampshire is fairly middle-of-the-road to conservative, and if we're concerned about this, then maybe you guys should pay attention to it."

New Hampshire towns last focused attention on a global issue during 1983, the issue being acid rain, a major issue in a state that is often upwind from major industries. "It might be somewhat normal for a [single] town to take on a national initiative," said Steve Norton, executive director of the New Hampshire Center for Public Policy Studies, "but not half the towns in the state" (Zezima 2007). In Bartlett, a town of about 2,200 in the White Mountains, the climate-change measure passed almost unanimously on March 15, 2007. The town's ski areas have been suffering from a lack of snow and long spells of above-freezing temperatures.

## WITH GOD ON EARTH'S SIDE

Religious people have been turning their attention more frequently and avidly toward environmental issues under the rubric of "creation care" or, as the Bible phrases it, "stewardship of the Earth." Large, drafty churches with high ceilings have been adding insulation through Interfaith Power and Light. The National Association of Evangelicals' vice president bought a Toyota Prius as more than fifty other evangelical Christian leaders pledged to neutralize their "carbon footprints" through energy conservation (Cooperman 2007b, A-1). Stewardship of the Earth is being applied to various types of greenhouse-gas-creating activities. Witness Leslie Lowe, energy and environmental director of the Interfaith Center on Corporate Responsibility, who characterizes the building of coal-fired power plants as, "under any faith-based view of the world, a blasphemy and a sin. It is to profit from the destruction of the natural world … and from the poisoning of other human life on this planet" (Jordan 2007b, D-1).

During 2007, evangelical Christian leaders openly split over the issue of global warming as a spiritual issue. Repudiating Christian radio commentator James C. Dobson, the board of directors of the National Association of Evangelicals said that environmental protection, which it calls "creation care," is a crucial moral issue. Dobson, founder of Focus on the Family, joining several other right-wing Christian leaders (including Gary L. Bauer, Tony Perkins, and Paul M. Weyrich), sent the board a letter denouncing the association's vice president, the Rev. Richard Cizik, for urging attention to global warming (Cooperman 2007a, A-5). The letter asserted that evangelicals were divided regarding whether climate change is a moral problem requiring their attention. Some right-wing religious leaders believed that making global warming a priority would divert resources from issues that they regard as more important, such as campaigns to ban abortions and same-sex marriages.

For some religious orders, creation care involves investment managers of theologically based assets, many of them quite small, who purchase small blocks of stock in energy and utility companies so that they can raise global warming as a moral issue at annual meetings. Patricia Daly, for example, who manages finances for a Catholic order, Sisters of Saint Dominic, of Caldwell, New Jersey, purchased 300 shares of ExxonMobil (of 5.5 billion shares outstanding), the world's largest and most profitable energy company, for the express purpose of introducing shareholder resolutions on the subject. Daly and other Catholic activists began filing global-warming resolutions in 1997, usually asking the company to account for its greenhouse-gas emissions. The first such resolution was approved by only 4.6 percent of proxy votes that year. By 2007, Daly's resolution won approval by 31 percent, or about 1.4 billion shares (Slater 2007, 22–27).

Meanwhile, in Europe, the Vatican during 2007 agreed to fund the planting of trees on 37 acres of weedy land near the Tisza River in Hungary in an effort to offset its carbon emissions. "As the Holy Father, Pope Benedict XVI, recently stated, the international community needs to respect and encourage a 'green culture,'" said Cardinal Paul Poupard, leader of the Pontifical Council for Culture. "The Book of Genesis tells us of a beginning in which God placed man as guardian over the earth to make it fruitful" (Rosenthal 2007). This is a new twist on the biblical passage describing Adam and Eve's expulsion from the Garden of Eden to go forth, multiply, and subdue the Earth.

The Vatican, which also has installed solar panels atop ancient buildings, has not yet embraced reduction of population as a strategy to implement its new green consciousness, however. After the Vatican agreement was announced, Msgr. Melchor Sánchez de Toca Alameda, an official at the Council for Culture at the Vatican, told the Catholic News Service that buying credits was like doing penance. "One can emit less $CO_2$ by not using heating and not driving a car, or one can do penance by intervening to offset emissions, in this case by planting trees," he said (Rosenthal 2007).

## REFERENCES

Abboud, Leila. 2007. "How Denmark Paved Way to Energy Independence." *Wall Street Journal*, April 16, A-1, A-13.

Associated Press. 2007b. "Bush Calls for Global Goals on Emissions." *New York Times*, May 31. http://www.nytimes.com/aponline/us/AP-Bush.html.

———. 2007c. "Court Rebukes Administration in Global Warming Case." *New York Times*, April 2. http://www.nytimes.com/aponline/business/AP-Scotus-Greenhouse-Gase.html.

———. 2007e. "Live Earth to Kick Off in Australia." *New York Times*, July 6. http://www.nytimes.com/aponline/world/AP-Music-Live-Earth.html.

Bennhold, Katrin. 2007a. "France Tells U.S. to Sign Climate Pacts or Face Tax." *New York Times*, February 1. http://www.nytimes.com/2007/02/01/world/europe/01climate.html.

Broder, John M. 2007. "Compromise Measure Aims to Limit Global Warming." *New York Times*, July 11. http://www.nytimes.com/2007/07/11/washington/11climate.htm.

Carlton, Jim. 2007. "Citicorp Tries Banking on the Natural Kind of Green." *Wall Street Journal*, September 5, B-1, B-8.

Chameides, William, and Michael Oppenheimer. 2007. "Carbon Trading over Taxes." *Science* 315 (March 23): 1670.

Cline, William R. 1992. *The Economics of Global Warming*. Washington, D.C.: Institute for International Economics.

Cooper, Glenda. 2007. "Live Earth London's Glacial Pacing: Mixing Music and a Serious Message Gives Concert a Clunky Rhythm." *Washington Post*, July 8, D-4. http://www.washingtonpost.com/wp-dyn/content/article/2007/07/07/AR2007070701201_pf.html.

Cooperman, Alan. 2007a. "Evangelical Body Stays Course on Warming; Conservatives Oppose Stance." *Washington Post*, March 11, A-5. http://www.washingtonpost.com/wp-dyn/content/article/2007/03/10/AR2007031001175_pf.html.

———. 2007b. "Eco-Kosher Movement Aims to Heed Tradition, Conscience." *Washington Post*, July 7, A-1. http://www.washingtonpost.com/wp-dyn/content/article/2007/07/06/AR2007070602092_pf.html.

Cortese, Amy. 2007. "Friend of Nature? Let's See Those Shoes." *New York Times*, March 6. http://www.nytimes.com/2007/03/06/business/businessspecial2/07label-sub.html.

Cowell, Alan. 2007. "Britain Drafts Laws to Slash Carbon Emissions." *New York Times*, March 14. http://www.nytimes.com/2007/03/14/world/europe/14britain.html.

Cox News Service. 2007. "Investors Sizing Up 'Carbon Footprints.'" *Omaha World-Herald*, September 30, D-1.

Deutsch, Claudia H. 2007a. "Companies Giving Green an Office." *New York Times*, July 3. http://www.nytimes.com/2007/07/03/business/03sustain.html.

Economists' Letter on Global Warming. 1997. June 23. http://uneco.org/global_warming.html.

Eilperin, Juliet. 2007b. "150 Global Firms Seek Mandatory Cuts in Greenhouse Gas Emissions." *Washington Post*, November 30, A-3. http://www.washingtonpost.com/wp-dyn/content/article/2007/11/29/AR2007112902039_pf.html.

Environmental Defense. 2007. "More CEOs Call for Climate Action." *Solutions* 38, no. 5 (November): 9. http://www.environmentaldefense.org.

Environment News Service. 2006c. "Pearl Jam Offsets Climate Footprint of 2006 World Tour." July 10.

———. 2007c. "Big City Mayors Strategize to Beat Global Warming." May 15. http://www.ens-newswire.com/ens/may2007/2007-05-15-01.asp.

———. 2007g. "Business Leaders Urge Climate Action." February 22. http://www.ens-newswire.com/ens/feb2007/2007-02-22-01.asp.

———. 2007i. "China Now Number One in Carbon Emissions; USA Number Two." June 19. http://www.ens-newswire.com/ens/jun2007/2007-06-19-04.asp.

———. 2007r. "Guide Delivers Live Earth Global Warming Survival Skills." June 19. http://www.ens-newswire.com/ens/jun2007/2007-06-19-02.asp.

———. 2007s. "House Approves 25 Percent Renewable Energy by 2025." October 17. http://www.ens-newswire.com/ens/oct2007/2007-10-17-091.asp.

———. 2007w. "Mandatory U.S. Greenhouse Gas Cap Wins New Corporate Supporters." May 8. http://www.ens-newswire.com/ens/may2007/2007-05-08-01.asp.

———. 2007y. "States Form Climate Registry to Track Greenhouse Gas Emissions." May 8. http://www.ens-newswire.com/ens/may2007/2007-05-08-09.asp#anchor1.

———. 2007aa. "U.S. Mayors Seek Federal Help to Protect Climate." November 5. http://www.ens-newswire.com/ens/nov2007/2007-11-05-01.asp.

———. 2007bb. "U.S. Mayors Take the Lead in Fighting Climate Change." June 25. http://www.ens-newswire.com/ens/jun2007/2007-06-25-04.asp.

———. 2007dd. "WWF: Stopping Climate Change Is Possible." May 2. http://www.ens-newswire.com/ens/may2007/2007-05-02-01.asp.

Fahrenthold, David A., and Steven Mufson. 2007. "Cost of Saving the Climate Meets Real-World Hurdles." *Washington Post*, August 16, A-1. http://www.washingtonpost.com/wp-dyn/content/article/2007/08/15/AR2007081502432_pf.html.

Faiola, Anthony, and Robin Shulman. 2007. "Cities Take Lead on Environment as Debate Drags at Federal Level." *Washington Post*, June 9, A-1. http://www.washingtonpost.com/wp-dyn/content/article/2007/06/08/AR2007060802779_pf.html.

Fairfield, Hannah. 2007. "When Carbon Is Currency." *New York Times*, May 6. http://www.nytimes.com/2007/05/06/business/yourmoney/06emit2.html.

Fialka, John J. 2007a. "Carbon Curbs Gain Backers." *Wall Street Journal*, February 27, A-8.

Frank, Robert. 2007. "Living Large While Being Green." Wealth Report. *Wall Street Journal*, August 24, W-2.

Gore, Albert, Jr. 1992. *Earth in the Balance: Ecology and the Human Spirit*. Boston: Houghton-Mifflin.

Hansen, James E. 2007b. "Political Interference with Government Climate Change Science." Testimony before the House Committee on Oversight and Government Reform, 110th Cong., 1st sess., March 19.

Hillman, Mayer, Tina Fawcett, and Sudhir Chella Rajan. 2007. *The Suicidal Planet: How to Prevent Global Climate Catastrophe*. New York: St. Martin's/Dunne.

Illinois Environmental Protection Agency. N.d. "Energy and Equity: The Full Report." http://www.cnt.org/ce/energy&equity.htm.

Jordan, Steve. 2007b. "Coal-Emission Cleanup a Challenge for Utilities." *Omaha World-Herald*, October 8, D-1, D-2.

Kelley, Kate. 2006. "City Approves 'Carbon Tax' in Effort to Reduce Gas Emissions." *New York Times*, November 18. http://www.nytimes.com/2006/11/18/us/18carbon.html.

Kher, Unmesh. 2007. "Pay for Your Carbon Sins." *Time*, April 9. http://www.time.com/time/printout/0,8816,1603737,00.html.

Kurtz, Howard. 2007. "RM + WSJ: Let's Do the Math." *Washington Post*, May 14, C-1.

Layton, Lyndsey. 2007. "A Carbon-Neutral House? Plan Would Offset Emissions by End of Current Congress." *Washington Post*, May 25, A-17. http://www.washingtonpost.com/wp-dyn/content/article/2007/05/24/AR2007052402146.html.

Mankiw, N. Gregory. 2007. "One Answer to Global Warming: A New Tax." *New York Times*, September 16. http://www.nytimes.com/2007/09/16/business/16view.html.

Martin, Andrew. 2007. "In Eco-friendly Factory, Low-Guilt Potato Chips." *New York Times*, November 15, A-1, A-22.

Masson, Gordon. 2003. "Eco-friendly Movement Growing in Music Biz." *Billboard*, March 15, 1.

Mufson, Steven. 2007a. "Federal Loans for Coal Plants Clash with Carbon Cuts." *Washington Post*, May 14, A-1. http://www.washingtonpost.com/wp-dyn/content/article/2007/05/13/AR2007051301105.html.

———. 2007b. "Florida's Governor to Limit Emissions." *Washington Post*, July 12, D-1. http://www.washingtonpost.com/wp-dyn/content/article/2007/07/11/AR2007071102139_pf.html.

Nordhaus, William D. 1991. "Economic Approaches to Greenhouse Warming." In *Global Warming: Economic Policy Reponses*, ed. Rudiger Dornbusch and James M. Poterba, 33–66. Cambridge, Mass.: MIT Press.

Pegg, J. R. 2007a. "U.S. Lawmakers Hear Stern Warnings on Climate Change." Environment News Service, February 14. http://www.ens-newswire.com/ens/feb2007/2007-02-14-10.asp.

———. 2007b. "U.S. Senators Propose Compulsory Greenhouse Gas Cuts." Environment News Service, October 18.

Revkin, Andrew. 2007. "Carbon-Neutral Is Hip, but Is It Green?" *New York Times*, April 29. http://www.nytimes.com/2007/04/29/weekinreview/29revkin.html.

Riding, Alan. 2007. "Stars Join Their Voices to Support Live Earth." *New York Times*, July 8. http://www.nytimes.com/2007/07/08/arts/music/08liveearth.html.

Rosenthal, Elisabeth. 2007. "Vatican Penance: Forgive Us Our Carbon Output." *New York Times*, September 17. http://www.nytimes.com/2007/09/17/world/europe/17carbon.html.

Samuelson, Robert J. 2007. "Prius Politics." *Washington Post*, July 25, A-15. http://www.washingtonpost.com/wp-dyn/content/article/2007/07/24/AR 2007072401855_pf.html.

Schwarzenegger, Arnold, and Jodi Rell. 2007. "States Can't Wait on Global Warming; Lead or Step Aside, EPA." *Washington Post*, May 21, A-13. http://www.washingtonpost.com/wp-dyn/content/article/2007/05/20/AR 2007052001059_pf.html.

Sisario, Ben. 2007. "Songs for an Overheated Planet." *New York Times*, July 6. http://www.nytimes.com/2007/07/06/arts/music/06eart.html.

Slater, Dashka. 2007. "Public Corporations Shall Take Us Seriously." *New York Times Sunday Magazine*, August 12, 22–27.

Solomon, Deborah. 2007. "Carbon's New Battleground: Lawmakers Favor Cap-and-Trade System, Economists Prefer a Tax." *Wall Street Journal*, September 12, A-4.

Spencer, Jane. 2007. "Big Firms to Press Suppliers on Climate." *Wall Street Journal*, October 9, A-7.

Walsh, Bryan. 2007a. "Get a Carbon Budget." *Time*, April 9. http://www.time.com/time/printout/0,8816,1603651,00.html.

Williams, Alex. 2007. "Buying into the Green Movement." *New York Times*, July 1. http://www.nytimes.com/2007/07/01/fashion/01green.html.

Zezima, Katie. 2007. "In New Hampshire, Towns Put Climate on the Agenda." *New York Times*, March 19. http://www.nytimes.com/2007/03/19/us/ 19climate.htm.

# CHAPTER 9

# Technofixes

Since the days when opposable thumbs, free arms, and brains first earned humankind a spot at the top of the food chain, we have been technological animals. The production of greenhouse gases themselves are the fruit of a machine age, the application of technology to increase profit, comfort, and convenience. Come now solutions that address technological problems with anthropomorphic solutions—filling the atmosphere with sulfur to counter the effects of greenhouse gases, erection of giant mirrors in the sky, synthesis of life that will consume carbon dioxide, grand plans on a grand scale. Many of these proposals seem to reveal a certain desperation on the part of a greenhouse-gas-addicted industrial system; each also runs up against practical limits of nature, raising a basic question: Given the fact that too much carbon dioxide, methane, and various other trace gases are the pollutants, isn't the best course to avoid making so much of them in the first place? Do we really want to fill the sky with sulfur dioxide so that we can continue to drive the gas hogs we treasure?

If it's too warm, suppose we erect a big umbrella to create more shade, cool the planet, and do it on a huge scale? Could we then continue to jack up atmospheric levels of greenhouse gases without worry? J. Roger P. Angel, a physicist at the University of Arizona, has suggested that billions of small lenses, each about two feet long, be launched into permanent orbit around the Earth to create a "solar sunshade" that could reflect 10 percent of the sun's light (Broad 2006). This solution would replace, though technological means, some of the bright, reflective surface that once was provided naturally by reflective ice. Before you fire up that SUV and figure that the global-warming mission has been accomplished, however, consider the sticker shock of this so-called solution: several trillion dollars over a quarter of a century (Fitzgerald 2007, 3).

In the meantime, coal-burning utilities in the United States are facing emissions curbs with the complicated, time-consuming, and expensive prospect of finding technological means to clean up their plants' effluents, the dirtiest of fossil fuels. This is no small problem, since almost half the electricity consumed in the United States is produced from coal. Projects are under way to turn coal into cleaner burning gas, meanwhile pumping unwanted carbon into the Earth's atmosphere or oceans. The price tag is in the billions of dollars, and many utilities are coming to relish a novel marketing strategy: the best kilowatt is the one that nobody uses.

## SULFUR AS SAVIOR: COOLING OFF
## IN AN ACID BATH

Some climate contrarians (and even some well-known scientists) have considered combating one kind of pollution—greenhouse gases—with another: sulfur. Air pollution from emissions of sulfur dioxide and other aerosols increase the net albedo (reflectivity) of the Earth, thus exerting a cooling influence on planetary temperatures.

Yet another proposal along the same lines involves burning sulfur in ships and power plants to form sulfate aerosols, or particles, to stimulate the formation of sunlight-reflecting clouds over the oceans. Mikhail Budyko, a Russian climatologist, has proposed a massive atmospheric infusion of sulfur that would form enough sulfur dioxide to wrap the Earth in thin, white, radiation-deflecting clouds within months. The net effect, according to Budyko, would be to cool the Earth in a fashion similar to the massive eruption of the volcano Tambora in 1815, which ejected enough sulfur into the air to produce, in 1816, an annual cycle known to climate historians as "the year without a summer." Crops across New England and upstate New York were devastated by frosts that continued into the late spring. Mohawk Indians at Akwesasne, in far northern New York, reported frosts into June that year.

Such atmospheric modification probably will remain an intellectual parlor game because the environmental costs of filling the stratosphere with sulfur dioxide (or other pollutants) far outweigh the benefits, even in a warmer and more humid world. The sulfur dioxide would have to be refreshed at least twice a month, as the previous load would have washed onto the Earth, planetwide, as acid rain. Sulfuric acid also tends to attract chlorine atoms, creating a chemical combination that could assist chlorofluorocarbons (CFCs) in devouring stratospheric ozone.

This debate raises an intriguing (if that is the word) question: Do we really want to intentionally poison the atmosphere so that we can continue to overburden it with greenhouse gases? Are we so far removed from nature that we would use one form of atmospheric pollution to counter the effects of another?

Mount Pinatubo in the Philippines erupted in 1991, ejecting about 10 million tons of sulfur into the atmosphere, enough to cool the Earth's near-surface atmosphere about half a degree for about a year or two, equaling (for a few months, anyway) the increase attributable to global

**Figure 9.1. Ash plume of Pinatubo during 1991 eruption, injecting sulfur into the atmosphere.**
Source: USGS.

warming during the previous century. The physical challenge of lifting that much sulfur into the atmosphere year after year would be quite a challenge, especially as increases in carbon dioxide levels over time require more of it.

Aside from the obvious pollution of the atmosphere, this "solution" has several problems. For example, most sulfur compounds leave the atmosphere within two weeks of their generation, while atmospheric carbon dioxide remains for a century or more. To negate the effects of global warming, the "pollution solution" would require a continuing feed of sulfur into the atmosphere. As has been noted above, the resulting precipitation of sulfur-enhanced acidity could have disastrous environmental effects at ground level.

Tom Wigley of the National Center for Atmospheric Research (NCAR) raises another problem: increasing acidity of the oceans, which already is becoming a threat to life, even absent intentional, additional human-caused contribution:

Projected anthropogenic warming and increases in $CO_2$ concentration present a two-fold threat, both from climate changes and from $CO_2$ directly through increasing the acidity of the oceans. Future climate change may be reduced through mitigation (reductions in greenhouse gas emissions) or through geoengineering. Most geoengineering approaches, however, do not address the problem of increasing ocean acidity. A combined mitigation/geoengineering strategy could remove this deficiency. Here we consider the deliberate injection of sulfate aerosol precursors into the stratosphere. This action could substantially offset future warming and provide additional time to reduce human dependence on fossil fuels and stabilize $CO_2$ concentrations cost-effectively at an acceptable level. (Wigley 2006, 452)

Pumping the atmosphere full of sulfur could have yet another side effect besides acid rain: it could also cause catastrophic drought. Kevin Trenberth and Aiguo Dai of NCAR have released a study in *Geophysical Research Letters* indicating that the 1991 eruption of Mount Pinatubo not only pumped vast amounts of sulfates into the atmosphere, reducing global temperatures by a few tenths of a degree for several years, but also inhibited the hydrological cycle worldwide, resulting in less precipitation and runoff (*New Scientist* 2007b, 16).

Despite evident problems, scientists have tried to treat the idea of seeding the atmosphere with sulfur dispassionately. "We should treat these ideas like any other research and get into the mindset of taking them seriously," said Ralph J. Cicerone, president of the National Academy of Sciences (Broad 2006).

Paul J. Crutzen of the Max Planck Institute for Chemistry in Mainz, Germany, who won a Nobel prize in 1995 for showing how industrial gases damage the Earth's stratospheric ozone shield, has examined the risks and benefits of trying to cool the planet by injecting sulfur into the stratosphere (Crutzen 2006). Many scientists discuss the sulfur solution with an air of sadness. Crutzen, for one, has written: "The very best [result] would be if emissions of the greenhouse gases could be reduced so much that the [geoengineering] experiment would not have to take place. Currently, this looks like a pious wish" (Kintisch 2007c, 1054). This idea "should not be taken as a license to go out and pollute," Cicerone said, as he emphasized that most scientists believe that curbing greenhouse gases should be the top priority. He added, however, that Crutzen "has written a brilliant paper" (Broad 2006).

In his paper, published along with several commentaries in the August 2006 issue of *Climatic Change*, Crutzen estimated the annual cost of injecting sulfur into the atmosphere at up to $50 billion, or about 5 percent of the world's annual military spending. "Climatic engineering, such as presented here, is the only option available to rapidly reduce temperature rises" if international efforts fail to curb greenhouse gases, Crutzen wrote. "So far," he added, "there is little reason to be optimistic" (Broad 2006). Crutzen cited the "grossly disappointing international political response" to increasing evidence of global warming (Kerr 2006, 401).

Some scientists object to such a scheme on grounds that it would reduce pressure to deal with the problem at its source—that is, to find energy sources other than fossil fuels. "I refuse to go down that road," said biochemist Meinrat Andreae of the Max Planck Institute. "You're papering over the problem so people can keep inflicting damage on the climate system" (Kerr 2006, 403). Depending on intentional pollution to counter increasing emissions of greenhouse gases is "like a junkie figuring out new ways of stealing from his children," said Meinrat (Morton 2007, 133).

"The biggest risk of geoengineering is that it eliminates pressure to decrease greenhouse gases," said Kenneth Caldeira of the Carnegie Institution Department of Global Ecology at Stanford University (Kerr 2006). Others argue that such a system could make the Earth dependent on a continuing human-provided sulfur "fix"—should the supply of sulfur falter. The Earth could heat up beyond sustainable levels within a few years.

Regarding geoengineering, James E. Hansen has said:

> I doubt the feasibility [and] desirability of geo-engineering suggestions such as the human volcano or space mirrors. If it proves necessary to counteract past emissions, why not a more "natural" method of drawing carbon dioxide out of the atmosphere: negative-carbon dioxide power plants that burn biofuels (derived, e.g., from cellulosic plants)? (Hansen, personal communication, February 27, 2007)

## THE MOON DUST SOLUTION

Could we ameliorate global warming with a giant dust cloud positioned in space to shade the sun? This sun shield, created with dust mined on the moon, has been proposed by Curtis Struck (2007) of Iowa State University.

Lunar dust particles are just the right size to scatter sunlight, Struck says. According to a piece in *New Scientist*, "If the particles are injected at two precise positions along the moon's orbit, they will form a pair of stable clouds that would each pass in front of the sun once a month, blocking sunlight for about 20 hours each month" (*New Scientist* 2007a). "If we're facing real threats to civilization, then you might resort to this sort of thing," said Struck. Others have criticized the scheme on the grounds that the clouds may reflect extra light onto Earth during the periods when they are not directly in front of the sun (*New Scientist* 2007a).

## "CLEAN" COAL-FIRED POWER: CARBON DIOXIDE SEQUESTRATION TECHNOLOGY

The electric power industry is just beginning to formulate practical plans to reduce its greenhouse-gas emissions, realizing that 80 percent or more emission cuts necessary to seriously curtail global warming in coming decades will require "completely remaking the electric industry," said Jim Dooley, senior staff scientist at the U.S. Department of Energy's (DOE) Pacific Northwest National Laboratory in Richland, Washington (Smith 2007a, A-10).

American Electric Power (AEP), with twenty-five large coal-powered electrical plants producing three-quarters of its power in eleven states, is the number-one corporate source of carbon dioxide in the United States. It is working on a coal-gasification plant in West Virginia that will cost at least $2.4 billion, enough by itself to raise electric rates in that state at least 12 percent. The reward is that the liquefied coal will produce less carbon dioxide. The plant may open and producing as much as 630 megawatts of power in 2012.

AEP also is working on the 600-acre Mountaineer coal-sequestration facility, also in West Virginia, which features the 9,172-foot "well to hell," into which the company during 2008 began pumping carbon dioxide through porous rock. The gas is held in place (hopefully for thousands of years) by a capstone. Before it is pumped into the Earth, some of the carbon dioxide is scrubbed out, a process which itself involves burning a lot of coal. Research is under way to make scrubbing more efficient. Under plans made public in 2007, gasification and sequestration will not be industrial-scale until about 2020.

An alliance of power companies, including AEP, was working under a $1.5 billion DOE grant to build a coal-fired power plant that would have produced no greenhouse-gas emissions. In January, 2008, the Energy Department canceled its 74 percent cost share in the "FutureGen" project, citing soaring costs and lack of results. The technology is untested, and the delivery date is uncertain—"in a decade," perhaps (Smith 2007a, A-10). In the meantime, General Electric has completed engineering work for a coal-fired power plant that (in theory) may operate with negligible greenhouse-gas emissions. According to one report, "General Electric also recently engineered jet engines, power turbines, and diesel locomotives that require less fuel and hence release less greenhouse gas, than those now in use" (Easterbrook 2007, 63).

Many technologies have been developed to remove carbon dioxide from the emissions of coal-fired plants. One popular approach is the Integrated Gasification Combined Cycle (IGCC), which creates hydrogen and $CO_2$ to be sequestered. Some technologies remove carbon dioxide from the flue stream after combustion. All of these technologies (and others) require energy and cost money and, as a result, would probably raise electric bills 40 to 50 percent, barring an unforeseen technological breakthrough.

Vattenfall, a Swedish power company, opened a carbon capture and storage plant in Swartze Pumpe, Germany, during 2008, claiming bragging rights for producing the world's first carbon-free coal-fired electricity. The 30-megawatt plant uses an "oxyfuel" combustion system that mixes oxygen with recycled flue gas containing carbon dioxide which, when burned, produces water and carbon dioxide. The waste $CO_2$ is stored underground, mainly in depleted oil and natural gas fields.

One experimental approach at the Warrior Run power plant in Cumberland, Maryland, captures carbon dioxide from its boiler and sells it to beverage gas distributors for use mainly in soft drinks. "If you've had a Coke today, you've probably ingested some of our product," said plant manager Larry Cantrell. The problem is that 4 megawatts of the plant's output (about 200 megawatts) goes to remove 5 percent of its carbon dioxide emissions (Kintisch 2007b, 185).

The best way to make additional coal-fired plants unnecessary would be to decrease or eliminate electrical demand. A considerable amount of energy can be saved merely by retrofitting existing buildings. We need to make a hard look at every kilowatt we use. A great deal of outdoor lighting, for example, could be dimmed or eliminated. Calgary, Alberta, for example, reduced its municipal electrical bill $2 million a year by dimming or eliminating street lights. The California Department of Transportation gives preference on highways to reflectors and other passive guides over streetlights. The safety value of floodlighting is often overrated. Full light merely gives burglars and vandals room in which to do their work. Motion-sensor lights are much better for capturing activity. The San Antonio public school system switched off lights in its buildings and parking lots at night and found that vandalism fell sharply (Owen 2007, 30).

## OCEAN SEQUESTRATION OF CARBON DIOXIDE

The oceans are well known to scientists as a major "sink," or repository, for atmospheric carbon dioxide and methane. This fact has led to various proposals to remove the nettlesome oversupply of these gases from the atmosphere and inject at least some of it into the depths of the oceans.

Peter G. Brewer, Gernot Friederich, Edward T. Peltzer, and Franklin M. Orr Jr. have demonstrated that deep-ocean disposal of carbon dioxide is technologically feasible. They described in *Science* a series of experiments during which carbon dioxide was lowered into seawater at different depths. The carbon dioxide, which is a gas at the surface, formed solid hydrates that were expected to remain in the ocean depths for "quite long residence times" (Brewer et al. 1999).

The same idea has been dissected and dismissed by Hein J. W. de Baar of the Netherlands Institute for Ocean Sciences:

> The crucial problem with fossil fuel $CO_2$ is its very rapid introduction within 100 to 200 years into the atmosphere, as opposed to the very slow response of many thousands to millions of years of the deep ocean in absorbing such $CO_2$. Eventually, the capacity for storage of $CO_2$ in the deep ocean is very large. Yet, in the meantime, we will witness a transient peak of atmospheric $CO_2$ which may yield catastrophic changes in the climate. Only after several thousands to millions of years most, but not all, of the fossil fuel $CO_2$ will be taken up by the oceans. (Baar and Stoll 1992, 143)

In addition, according to Baar, deep-sea carbon dioxide injection is suitable only for large stationary energy plants (30 percent of total human emissions) and would raise their cost of generation 30 to 45 percent, while decreasing efficiency by a similar percentage. Much of the carbon dioxide injected into the deep oceans would eventually return to the surface, doubling seawater's acidity, which would be toxic for fish, plankton, and other life in the oceans. "Deep-sea injection is at best a partial, expensive, and [temporary] remedy to the $CO_2$ problem," wrote Baar (Baar and Stoll 1992, 144).

The oceans' capacity to absorb excess carbon is hardly limitless. As with the "sulfur solution," deep-sea sequestration of carbon dioxide could further acidify the oceans. Most proposals to inject human-generated carbon dioxide into the oceans ignore the possible effects of such sequestration on marine life. Brad A. Seibel and Patrick J. Walsh examined these effects, finding that increased deepwater carbon dioxide levels result in decreases of seawater pH (increasing acidity), which can be harmful to sea creatures, "as has been demonstrated for the effects of acid rain on freshwater fish" (Seibel and Walsh 2001, 319). They find that "a drop in arterial pH by just 0.2 would reduce bound oxygen in the deep-sea crustacean *Glyphocrangon vicaria* by 25 percent" (2001, 320). The same drop in arterial pH would reduce bound oxygen in the midwater shrimp *Gnathophausia ingens* by 50 percent. According to Seibel and Walsh, "Deep-sea fish hemoglobins are even more sensitive to pH" (2001, 320).

Small increases in carbon dioxide levels and resulting decreases of pH levels "may trigger metabolic suppression in a variety of organisms," as "low pH has been shown to inhibit protein synthesis in trout living in lakes rendered acidic through anthropogenic effects" (Seibel and Walsh 2001, 320). Seibel and Walsh cited research by R. L. Haedrich that "any change that takes place too quickly to allow for a compensating adaptive change within the genetic potential of finely adapted deep-water organisms is likely to be harmful" (Seibel and Walsh 2001, 320). They conclude: "Available data indicate that deep-sea organisms are highly sensitive to even modest pH changes.... Small perturbations in $CO_2$ or pH may thus have important consequences for the ecology of the deep sea" (2001, 320).

"Through various feedback mechanisms, the ocean circulation could change and affect the retention time of carbon dioxide injected into the deep ocean, thereby indirectly altering oceanic carbon storage and atmospheric carbon dioxide concentration," said Atul Jain, a professor of atmospheric sciences at the University of Illinois at Urbana-Champaign. "Where you inject the carbon dioxide turns out to be a very important issue" (Environment News Service 2002a). Jain and graduate student Long Cao found that climate change has an important impact on the oceans' ability to store carbon dioxide. The effect was most pronounced in the Atlantic Ocean. "When we ran the model without the climate feedback mechanisms, the Pacific Ocean held more carbon dioxide for a longer period of time," Cao said. "But when we added the feedback mechanisms, the retention time in the Atlantic Ocean proved far superior. Based on our initial results, injecting carbon dioxide into the Atlantic Ocean would be more effective than injecting it at the same depth in either the Pacific Ocean or the Indian Ocean" (Environment News Service 2002a).

Future climate change could affect both the uptake of carbon dioxide in the ocean basins and the ocean circulation patterns themselves, Jain said. "As sea-surface temperatures increase, the density of the water decreases and thus slows down the ocean Thermohaline Circulation, so the ocean's ability to absorb carbon dioxide also decreases," Jain explained. "This leaves more carbon dioxide in the atmosphere, exacerbating the problem. At the same time, the reduced ocean circulation will decrease the ocean mixing, which decreases the ventilation to the atmosphere of carbon injected into the deep ocean. Our model results show that this effect is more dominating in the Atlantic Ocean" (Environment News Service 2002a).

"Sequestering carbon in the deep ocean is, at best, a technique to buy time," Jain concluded. "Carbon dioxide dumped in the oceans won't stay there forever. Eventually it will percolate to the surface and into the atmosphere" (Environment News Service 2002a).

## OPPOSITION RISES TO CARBON DIOXIDE SEQUESTRATION

By 2002, tourism promoters, commercial fishermen, environmentalists, and sports groups had united in the Hawaiian Islands to oppose experiments in carbon dioxide sequestration off the island of Kauai. The effort has drawn support from Chevron/Texaco, General Motors, Ford, and ExxonMobil. Opponents fear that it will increase the acidity of the local ocean water and imperil animal and plant life. The DOE had allowed the companies to begin work without an environmental impact statement because it asserted that human beings would not be affected. Rep. Patsy Mink (D-Hawaii) said that "Hawaii's ocean environment is too precious to put at risk for an experiment of this kind" (Dunne 2002, 2).

The same company that tried to inject carbon dioxide into the sea off Hawaii also was lobbying to deposit 5.4 tons of pure $CO_2$ deep under the North Sea near Norway. This experiment was set to begin during the summer of 2002 until environmentalists campaigned successfully to stop it. According to an Environment News Service report:

> The Norwegian oil firm Statoil already was injecting roughly 1 million tons of $CO_2$ per year into the rock strata of an offshore oilfield in the North Sea, but no one has yet tried sequestration in the oceans. Led by the Norwegian Institute for Water Research, a coalition including United States, Japanese, Canadian and Australian organizations was planning to inject five tons of liquid $CO_2$ at 800 meters [2,600 feet] depth off the coast of Norway. (Environment News Service 2002c)

Norway's Pollution Control Agency granted the project a discharge permit in early July 2002, subject to approval by the Environment Ministry.

Environmental groups argued that the North Sea project would mean "dumping" $CO_2$ in the ocean in violation of the 1972 London Dumping Convention as well as the 1992 Ospar Convention on Protection of the North Sea Environment. The Ospar Commission discussed this issue in late June 2002 (Environment News Service 2002c). "The sea is not a dumping ground. It's illegal to dump nuclear or toxic waste at sea, and it's illegal to dump $CO_2$—the fossil fuel industry's waste," said Truls Gulowsen, a Greenpeace Norway climate campaigner (Environment News Service 2002c).

A last-minute veto from Norway's environment minister Borge Brende on August 26, 2002, stopped the project. "The possible future use of the ocean as a storage place for $CO_2$ is controversial" and "could violate current international rules concerning sea waters," said Brende (Agence France-Presse 2002).

## IRON FERTILIZATION OF THE OCEANS

Nearly half of the Earth's photosynthesis is performed by phytoplankton in the world's seas and oceans. This fact has led to proposals to "geoengineer" a carbon dioxide sink via the same "biological pump" that is believed to have driven at least some of the Earth's past climate cycles (Chisholm 2000, 685). Should the oceans be seeded with large amounts of iron ore that will stimulate the growth of carbon dioxide–consuming phytoplankton? The idea has attracted some support among corporations and foundations looking for ways to minimize the effects of carbon dioxide without changing the world's basic energy-generation mix. The idea is simple on its face: iron stimulates the growth of phytoplanktonic algae that are believed to be responsible for about half of the world's biologic absorption of carbon dioxide.

Ulf Riebesell, a marine biologist at the Alfred Wegener Institute for Polar and Marine Research in Bremerhaven, Germany, believes that ambitious iron seeding of the oceans could remove 3 to 5 billion tons of carbon dioxide per year, or about 10 to 20 percent of human-generated emissions. It has been said that such action would be at least ten times cheaper than planting forests that would remove the same amount of $CO_2$ (Schlermeier 2003, 110). Patents have been issued for ocean fertilization, and demonstration projects undertaken. One such project was described and evaluated in *Nature* (Boyd et al. 2000, 695–702; Watson et al. 2000, 730–33).

Scientists who fed tons of iron into the Southern Ocean reported evidence during 2004 that stimulating the growth of phytoplankton in this way may strengthen the oceans' use as a carbon sink. In a report in the April 16, 2004, issue of *Science*, ocean biologists and chemists from more than twenty research centers said they triggered two huge blooms of phytoplankton that turned the ocean green for weeks and consumed hundreds, perhaps thousands of tons of carbon dioxide. "These findings would be encouraging to those considering iron fertilization as a global geoengineering strategy," said Ken Coale, a chief scientist at the Moss Landing Marine Laboratories. "But the scientists involved in this experiment realize that this looked only skin deep at the functioning of ocean ecosystems and much more needs to be understood before we recommend such a strategy on a global scale" (Hoffman 2004).

The chemistry of the oceans varies widely with regard to the amount of iron necessary to prime this pump and substantially increase carbon dioxide sequestration. In the equatorial Pacific and Southern Ocean, Sallie W. Chisholm, a marine biologist at the Massachusetts Institute of Technology, wrote in *Nature* that it "is possible to stimulate the productivity of hundreds of square kilometers of ocean with a few barrels of fertilizer" (Chisholm 2000, 686).

In an experiment conducted between Tasmania and Antarctica, researchers confirmed that vast stretches of the world's southern waters are primed to explode with photosynthesis if iron is added. The researchers said, however, that it is too soon to start large-scale iron seeding because the new experiment raised as many questions as it answered. At best, they said, iron seeding would absorb only a small amount of the carbon dioxide in the atmosphere. They also said that their experimental bloom of plankton was not tracked long enough to determine whether the carbon harvested from the air sank into the deep sea or was again released into the environment as carbon dioxide gas. Atsushi Tsuda and colleagues have studied iron fertilization and have found that, under some circumstances, iron fertilization can dramatically increase phytoplankton mass (Tsuda et al. 2003, 958–61).

## "THERE IS NO FREE LUNCH"

The seemingly simple proposition of iron seeding has some potential problems, however. "There are some fundamental scientific questions that need to be addressed before anyone can responsibly promote iron fertilization as a climate control tactic," said Coale (Revkin 2000a, A-18). First, there exists no way to measure the amount of carbon taken up by phytoplankton. Additionally, the algae produce dimethylsulphide, which plays a role in cloud formation. Phytoplankton also increase the amount of sunlight absorbed by ocean water, as well as heat energy. They produce compounds such as methyl halides, which play a role in stratospheric ozone depletion. The iron also could promote the growth of toxic algae, which may kill other marine life and change the chemistry of ocean water by removing oxygen. "The oceans are a tightly linked system, one part of which cannot be changed without resonating through the whole system," said Chisholm. "There is no free lunch" (Schlermeier 2003, 110).

So much iron may be required to produce the desired effect that fertilization of this type will never become commercially viable, even assuming solutions are found for other problems. "The experiments enabled us to make an initial determination about the amount of iron that would be required and the size of the area to be fertilized," said Ken O. Buesseler of the Woods Hole Oceanographic Institution, who coauthored a study of the idea. "Based on the studies to date, the amount of iron needed and area of ocean that would be impacted is too large to support the commercial application of iron to the ocean as a solution to our greenhouse gas problem," he explained (Environment News Service 2003b). "It may not be an inexpensive or practical option if what we have seen to date is true in further experiments on larger scales over longer time spans," Buesseler said. "The oceans are already naturally taking up human-produced carbon dioxide, so the changes to the system are already under way," he said. "We need to first ask will it work and then what are the environmental consequences?" (Environment News Service 2003b).

One study asserted, "To assess whether iron fertilization has potential as an effective sequestration strategy, we need to measure the ratio of iron added to the amount of carbon sequestered in the form of particulate organic carbon to the deep ocean in field studies" (Buesseler and Boyd 2003, 67). To date, wrote Buesseler and Philip W. Boyd, experiments of this type have "produced notable increases in biomass and associated decreases in dissolved inorganic carbon and macronutrients. However, evidence of sinking carbon particle carrying P.O.C. [particulate organic carbon] to the deep ocean was limited" (2003, 67). Given the limits of present technology, this study estimates that an area a magnitude larger than the Southern Ocean (waters south of 50 degrees south latitude) would have to be fertilized to remove 30 percent of the carbon that human activity presently injects into the atmosphere. Thus, according to this study, "ocean iron fertilization may not be a cheap and attractive option if impacts on carbon export and sequestration are as low as observed to date" (Buesseler and Boyd 2003, 68). "From my work, I don't think this could solve a significant fraction of our greenhouse-gas problem while causing unknown ecological consequences," said Buesseler (Hoffman 2004).

## A POLYMER THAT ABSORBS CARBON DIOXIDE

According to a report by the Environment News Service, scientists at the DOE's Los Alamos National Laboratory (LANL) have been developing a new high-temperature polymer membrane that will separate and capture carbon dioxide before it reaches the atmosphere. This technology is aimed at sequestering at least some of the 30 percent of anthropogenic carbon dioxide resulting from electrical generation. Present technology is limited to dealing with waste carbon dioxide emitted by such plants up to 150°C, but the wastes often reach 375°C.

Speaking at an American Geophysical Union conference on May 29, 2002, in Washington D.C., Jennifer Young, principal investigator for LANL's carbon dioxide membrane separation project, described a new polymeric-metallic membrane that is stable at temperatures as high as 370°C. "Current technologies for separating carbon dioxide from other gases require that the gas stream be cooled to below 150 degrees Celsius, which reduces energy efficiency and increases the cost of separation and capture," said Young. "By making a membrane which functions at elevated temperatures, we increase the practicality and economic feasibility of using membranes in industrial settings" (Environment News Service 2002b).

## PUT A BOUNTY ON CARBON DIOXIDE

As momentum built to address global warming, one strategy offers prizes for new technology to remove greenhouse gases from the atmosphere. During February 2007, billionaire Richard Branson, head of Virgin Group, offered a $25 million prize for a system that will remove a billion tons of carbon dioxide per year from the atmosphere. Judges of the prize include former vice president Al Gore, James Hansen, and James Lovelock (Tierney 2007). *New York Times* columnist John Tierney remarked:

> If governments and other moguls throw in more money, the new Virgin Earth Challenge may be the start of competitions that ultimately yield

nanobots or microbes capable of gobbling up carbon dioxide. As far-fetched as it seems today, removing carbon dioxide from the atmosphere could turn out to be a lot more practical than the alternative: persuading six billion people to stop putting it there. (Tierney 2007)

Branson is spending millions of dollars a year on research at Virgin Fuels, a new firm with ambitions to produce carbon-neutral fuel for all vehicles, including aircraft. He has committed more than $400 million for research into the feasibility of using enzymes and genetically modified organisms to produce clean fuels. Branson's initial focus, however, has been on various forms of ethanol, including butanol.

Branson owns two small islands in the British Virgin Islands, one a resort that rents for $46,000 a day when he is elsewhere. Branson says he plans to use a combination of solar, wind, and tidal power to eliminate fossil-fuel use from the two islands. No plans were evident, however, regarding how he will replace the helicopters that ferry celebrity guests to and from the islands.

## CARBON DIOXIDE'S USE AS A REFRIGERANT

Several hundred researchers from around the world met at Purdue University between July 12 and 15, 2004, to discuss innovative air-conditioning and refrigeration technologies, including designs directed at reducing carbon dioxide levels in the atmosphere. Ironically (as if chemicals could understand irony!), one of the ideas under consideration is the use of carbon dioxide as a refrigerant. Carbon dioxide was the first refrigerant used during the early twentieth century, but was later replaced with synthetic chemicals. Carbon dioxide may be on the verge of a comeback because of technological advances that include the manufacture of extremely thin aluminum tubing called "microchannels" (AScribe Newswire 2004).

Hydrofluorocarbons (HFCs), today's most widely used refrigerants, have about 1,400 times the heat-retention capacity of the same quantity of carbon dioxide. The HFCs replaced chlorofluorocarbons (CFCs), which are also potent greenhouse gases that degrade the stratospheric ozone that protects life at the surface from carcinogenic ultraviolet radiation. In this case, carbon dioxide would cause less harm to the atmosphere than HFCs. Tiny quantities of carbon dioxide released from air conditioners would be insignificant compared to the huge amounts produced from burning fossil fuels for energy and transportation, said Eckhard Groll, an associate professor of mechanical engineering at Purdue (AScribe Newswire 2004).

Carbon dioxide offers few advantages for large air conditioners, which do not have space restrictions and can use wide-diameter tubes capable of carrying enough of the conventional refrigerants to provide proper cooling capacity. Carbon dioxide, however, may be a promising alternative for systems that must be small and lightweight, such as automotive or portable air conditioners. Various factors, including the high operating pressure required for carbon dioxide systems, enable the refrigerant to flow through small-diameter tubing, which allows engineers to design more compact air conditioners (AScribe Newswire 2004).

In the meantime, Greenpeace has asked refrigerator manufacturers to use hydrocarbons such as propane instead of HFCs, having developed its own "Greenfreeze" technology. Unilever became interested in this idea as part of its commitment to reduce the impact of its activities on climate change. While the Greenfreeze technology was adequate for domestic refrigerators, Unilever carried out lengthy trials to create a system that could be used in larger-scale freezers as well. Unilever agreed to trials at the 2000 Olympics in Sydney and committed to a global changeover to HFC-free refrigerants.

## CREATION OF AN ORGANISM THAT WILL CONSUME CARBON DIOXIDE

J. Craig Venter, who compiled a human genetic map with private money, has decided to tackle the problem of global warming with a $100 million research endowment created from his stock holdings. Venter plans to scour deep ocean trenches for bacteria that may convert carbon dioxide to solid form using very little sunlight or other energy (Gillis 2002, E-1). As he searches for organisms in nature, Venter also is working to build one himself—the first synthetic life. Ideally, Venter's organism or group of organisms, produced via synthetic biology, will be able to take in carbon dioxide, break it down, and produce both biological compounds and biofuel. In other words: eat carbon dioxide, excrete biofuel.

By 2007, Venter and colleagues had taken a basic step along this path by changing one species of bacteria into another, called *Mycoplasma laboratorium*, for which he was seeking a patent. A team led by Venter already had created a synthetic virus. Viruses cannot reproduce (they infect hosts to do that), so they are not considered to be alive. Venter's team is creating a genome for a new life form. The next step will be "to spark that synthetic genome into life, so that it becomes a full-functioning organism" (Naik 2007b, B-1).

On January 23, 2008, researchers led by Venter said that they had, for the first time, manufactured the entire genome of a tiny natural bacterium called *Mycoplasma* by stitching together its components. The researchers did not take the next and more crucial step, however: inserting a synthetic chromosome into a living microbe. George M. Church, a professor of genetics at Harvard Medical School, said, "Right now, all they've done is shown they can buy a bunch of DNA and put it together" (Pollack 2008).

Ideally, Venter would like to invent two synthetic microorganisms. One would consume carbon dioxide and turn it into raw materials, including the kinds of organic chemicals that are now made from oil and natural gas. The other microorganism would generate hydrogen fuel from water and sunshine. "Other groups are considering capturing carbon dioxide and pumping it down to the bottom of the ocean, which would be

**Figure 9.2. J. Craig Venter.**
Source: A New Human Genome Sequence Paves the Way for Individualized Genomics by
Liza Gross (September 2007). *Biology* Vol. 5, No. 10. Wikipedia Commons.


insane," Venter told the *Financial Times*. "Why risk irreversible damage
to the ocean when we could be doing something useful with the carbon?"
(Cookson and Firn 2002, 11).

"We've barely scratched the surface of the microbial world out there to
try to help the environment," Venter said. "We're going to be searching
for some dramatic new microbes" (Gillis 2002, E-1). Venter said that his
ventures will be established as not-for-profit corporations. "I'm not in
business anymore," he said (Gillis 2002, E-1). Venter, who calls his organization the Institute for Biological Energy Alternatives, will seek grant
money from DOE. His goal will be to explore whether modern science
can use the power of biology to solve the world's most serious environmental crisis.

Venter has proposed installing colonies of organisms in "bioreactors"
near power plants to consume emissions of carbon dioxide and turn the
gas into solids such as sugars, proteins, and starches, mimicking the behavior of green plants (Gillis 2002, E-1). According to an account in the
*Washington Post*:

> Venter plans to base his approach on one of the most striking developments
> in biology in recent years—the discovery, in deep ocean trenches and vol-
> canic hot spots on the ocean floor, of a wide array of bacteria that can per-
> form extensive chemical reactions without needing sunlight. These are
> thought to be descendants of the most primitive life forms that arose on the

Earth, and scientists are just beginning to explore their potential. (Gillis 2002, E-1)

## A TOKYO SEAWATER COOLING GRID

Tokyo has been building the world's largest cooling system filled with seawater in an attempt to reduce temperatures in the city by 2.9°C (5.2°F)—or about the amount of warming Tokyo experienced during the twentieth century. General warming is being enhanced in Tokyo by an intensifying urban heat-island effect as population density increases and more air conditioning (with its waste heat) is being used to cool tall buildings. This huge construction project, estimated to cost $300 billion, requires

> a lattice of pipes under 250 hectares (about 600 acres) of the city, drawing in cool water from Tokyo Bay. Heat will be forced into the network through a second set of pipes connected to air conditioning systems in the buildings above. The heat will be absorbed by the seawater, which would then flow back into the bay. (Ryall 2002, 20)

Skeptics of the project fear that dumping warmed seawater in Tokyo Bay could cause environmental damage.

Tokyo's average temperature rise during the twentieth century was five times the average worldwide rate of warming. The number of days when the city experiences temperatures of 30°C (86°F) or more doubled during the last two decades of the century (Ryall 2002, 20). "This is a radical approach but that's what the situation demands," said Tadafumi Maejima, director of the Japan District Heating and Cooling Association, which was commissioned to draw up the plan. "Nothing like this has been attempted anywhere in the world, but we could be ready to start work in as little as two years" (Ryall 2002, 20).

According to a report in the London *Times*:

> The pipes would initially spread beneath 123 hectares [304 acres] of the Marunouchi district, the city's financial heart. Later they would extend through Kasumigaseki, where the Diet (parliament), and the ministries are based. A further phase, covering an area north of Tokyo station, would use water from a sewage treatment plant. (Ryall 2002, 20)

"The amount of heat emitted by buildings needs to be cut, especially in the heart of the city where demand for air conditioners is concentrated, to break the vicious cycle of summer warming," said the association's report (Ryall 2002, 20).

## REFERENCES

Agence France-Presse. 2002. "Norway Says No to Controversial Plan to Store $CO_2$ on Ocean Floor." August 22.

AScribe Newswire. 2004. "Conferences Tackle Key Issues in Air Conditioning, Refrigeration." June 23 (in LEXIS).

Baar, Hein J. W. de, and Michel H. C. Stoll. 1992. "Storage of Carbon Dioxide in the Oceans." In *Arctic Ecosystems in a Changing Climate: An Ecophysiological Perspective*, ed. F. Stuart Chapin III, Robert L. Jefferies, James F. Reynolds, Gaius R. Shaver, and Josef Svoboda, 143–77. San Diego: Academic Press.

Boyd, Philip W., Andrew J. Watson, Cliff S. Law, Edward R. Abraham, Thomas Trull, Rob Murdoch, Dorothee C. E. Bakker, Andrew R. Bowie, K. O. Buesseler, Hoe Chang, Matthew Charette, Peter Croot, Ken Downing, Russell Frew, Mark Gall, Mark Hadfield, Julie Hall, Mike Harvey, Greg Jameson, Julie LaRoche, Malcolm Liddicoat, Roger Ling, Maria T. Maldonado, R. Michael McKay, Scott Nodder, Stu Pickmere, Rick Pridmore, Steve Rintoul, Karl Safi, Philip Sutton, Robert Strzepek, Kim Tanneberger, Suzanne Turner2, Anya Waite, and John Zeldis. 2000. "A Mesoscale Phytoplankton Bloom in the Polar Southern Ocean Stimulated by Iron Fertilization." *Nature* 407(October 12): 695–702.

Brewer, Peter G., Gernot Friederich, Edward T. Peltzer, and Franklin M. Orr Jr. 1999. "Direct Experiments on the Ocean Disposal of Fossil Fuel $CO_2$." *Science* 284 (May 7): 943–45.

Broad, William J. 2006. "How to Cool a Planet (Maybe)." *New York Times*, June 27. http://www.nytimes.com/2006/06/27/science/earth/27cool.html.

Buesseler, Ken O., and Philip W. Boyd. 2003. "Will Ocean Fertilization Work?" *Science* 300 (April 4): 67–68.

Chisholm, Sallie W. 2000. "Stirring Times in the Southern Ocean." *Nature* 407 (October 12): 685–86.

Cookson, Craig, and David Firn. 2002. "Breeding Bugs That May Help Save the World; Craig Venter Has Found a Large Project to Follow the Human Genome." *Financial Times* (London), September 28, 11.

Crutzen, Paul J. 2006. "Albedo Enhancement by Stratospheric Sulfur Injections: A Contribution to Resolve a Policy Dilemma?" *Climatic Change* 77, no. 3 (August).

Dunne, Nancy. 2002. "Climate Change Research Sparks Hawaii Protests." *Financial Times* (London), June 20, 2.

Easterbrook, Gregg. 2007. "Global Warming: Who Loses—and Who Wins?" *Atlantic*, April, 5–64.

Environment News Service. 2002a. "Global Warming Could Hamper Ocean Sequestration." December 4. http://ens-news.com/ens/dec2002/2002-12-04-09.asp.

———. 2002b. "Hot Polymer Catches Carbon Dioxide." May 29. http://www.ens-newswire.com/ens/may2002/2002-05-29-09.asp#anchor5.

———. 2002c. "Liquid $CO_2$ Dump in Norwegian Sea Called Illegal." July 11. http://ens-news.com/ens/jul2002/2002-07-11-02.asp.

———. 2003b. "Iron Link to $CO_2$ Reductions Weakened." April 10. http://ens-news.com/ens/apr2003/2003-04-10-09.asp#anchor8.

Fitzgerald, Michael. 2007. "It Takes Deep Pockets to Fight Global Warming." *New York Times*, August 12, Business, 3.

Gillis, Justin. 2002. "A New Outlet for Venter's Energy; Genome Maverick to Take On Global Warming." *Washington Post*, April 30, E-1.

Hoffman, Ian. 2004. "Iron Curtain over Global Warming; Ocean Experiment Suggests Phytoplankton May Cool Climate." *Daily Review* (Hayward, Calif.), April 17 (in LEXIS).

Kerr, Richard A. 2006. "Pollute the Planet for Climate's Sake?" *Science* 314 (October 20): 401–3.

Kintisch, Eli. 2007b. "Making Dirty Coal Plants Cleaner." *Science* 317 (July 13): 184–86.

———. 2007c. "Scientists Say Continued Warming Warrants Closer Look at Drastic Fixes." *Science* 318 (November 16): 1054–55.

Morton, Oliver. 2007. "Is This What It Takes to Save the World?" *Nature* 447 (May 10): 132–36.

Naik, Gautam. 2007b. "J. Craig Venter's Next Big Goal: Creating New Life." *Wall Street Journal*, June 29, B-1.

*New Scientist*. 2007a. "Keep Earth Cool with Moon Dust." February 10. http://environment.newscientist.com/channel/earth/mg19325905.300-keep-earth-cool-with-moon-dust.html.

———. 2007b. "Volcanic Approach to Solving Climate Woes Goes Up in Smoke." August 11, 16.

Owen, David. 2007. "The Dark Side: Making War on Light Pollution." *New Yorker*, August 20, 28–33.

Pollack, Andrew. "Scientists Take New Step Toward Man-Made Life." *New York Times*, January 24, 2008. http://www.nytimes.com/2008/01/24/science/24cnd-genome.html.

Revkin, Andrew C. 2000a. "Antarctic Test Raises Hope on a Global-Warming Gas." *New York Times*, October 12, A-18.

Ryall, Julian. 2002. "Tokyo Plans City Coolers to Beat Heat." *Times* (London), August 11, 20.

Schlermeier, Quirin. 2003. "The Oresmen." *Nature* 421 (January 9): 109–10.

Seibel, Brad A., and Patrick J. Walsh. 2001. "Potential Impacts of $CO_2$ Injection on Deep-Sea Biota." *Science* 294 (October 12): 319–20.

Smith, Rebecca. 2007a. "Inside Messy Reality of Cutting $CO_2$ Output." *Wall Street Journal*, July 12, A-1, A-10.

Struck, Curtis. 2007. "The Feasibility of Shading the Greenhouse with Dust Clouds at the Stable Lunar Lagrange Points." *Journal of the British Interplanetary Society* 60 (March): 82–89.

Tierney, John. 2007. "A Cool $25 Million for a Climate Backup Plan." *New York Times*, February 13. http://www.nytimes.com/2007/02/13/science/earth/13tier.html.

Tsuda, Atsushi, Shigenobu Takeda, Hiroaki Saito, Jun Nishioka, Yukihiro Nojiri, Isao Kudo, Hiroshi Kiyosawa, et al. 2003. "A Mesoscale Iron Enrichment in the Western Subarctic Pacific Induces a Large Centric Diatom Bloom." *Science* 300 (May 9): 958–61.

Watson, A. J., D. C. E. Bakker, A. J. Ridgwell, P. W. Boyd, and C. S. Law. 2000. "Effect of Iron Supply on Southern Ocean $CO_2$ Uptake and Implications for Glacial Atmospheric $CO_2$." *Nature* 407 (October 12): 730–33.

Wigley, T. M. L. 2006. "A Combined Mitigation/Geoengineering Approach to Climate Stabilization." *Science* 314 (October 20): 452–54.

# The Rationale for a Moratorium on New $CO_2$-Emitting Power Plants

## By James E. Hansen

## BASIC FOSSIL FUEL FACTS

The role of coal in global warming is clarified by a small number of well-documented facts. Figure 1 shows the fraction of fossil fuel carbon dioxide ($CO_2$) emissions that remains in the air over time. One-third of the $CO_2$ is still in the air after 100 years, and one-fifth is still in the air after 1000 years.

Oil slightly exceeds coal as a source of $CO_2$ emissions today, as shown in Figure 2a. [IPCC = Intergovernmental Panel on Climate Change; WEC = World Energy Council] But, because of the long atmospheric lifetime of past emissions, fully half of the excess $CO_2$ in the air today (from fossil fuels), relative to pre-industrial times, is from coal (Figure 2b). Moreover, coal use is now increasing, while oil production has stagnated. Oil production will peak and will be constrained by available resources earlier than will coal production.

Figure 3 shows reported fossil fuel reserves and resources (estimated undiscovered deposits). Reserves are hotly debated and may be exaggerated, but we know that enough oil and gas remain to take global warming close to, if not into, the realm of dangerous climate effects. Coal and unconventional fossil fuels such as tar shale contain enough carbon to produce a vastly different planet, a more dangerous and desolate planet, from the one on which civilization developed, a planet without Arctic sea ice, with crumbling ice sheets that ensure sea level catastrophes for our children and grandchildren, with shifting climate zones that cause great hardship for the world's poor and drive countless species to extinction, and

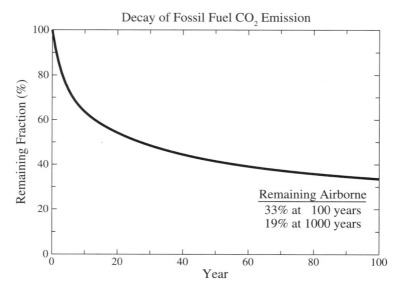

Figure 1. The fraction of $CO_2$ remaining in the air, after emission by fossil fuel burning, declines rapidly at first, but 1/3 remains in the air after a century and 1/5 after a millennium (Atmos. Chem. Phys. 7, 2287–2312, 2007).

with intensified hydrologic extremes that cause increased drought and wildfires but also stronger rain, floods, and storms.

Oil and coal uses differ fundamentally. Oil is burned primarily in small sources, in vehicles where it is impractical to capture the $CO_2$ emissions.

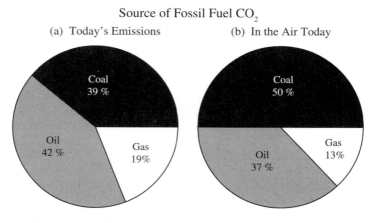

Figure 2. Percent contributions of different fossil fuels to 2006 $CO_2$ emissions (left side) and contributions to the excess $CO_2$ in the air today relative to pre-industrial $CO_2$ amount (CDIAC data for 1751–2004, BP for 2005–6; cf. Atmos. Chem. Phys. 7, 2287–2312, 2007).

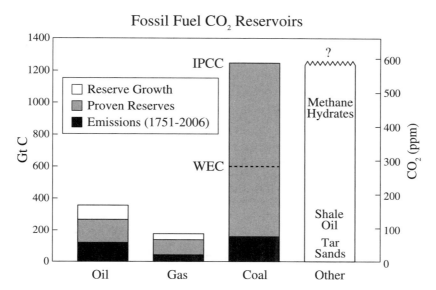

Figure 3. **Estimated fossil fuel reserves; darker portions have already been used (Atmos. Chem. Phys. 7, 2287–2312, 2007).**

Available oil reserves will be exploited eventually, regardless of efficiency standards on vehicles, and the $CO_2$ will be emitted to the atmosphere. The climate effect of oil is nearly independent of how fast we burn the oil, because much of the $CO_2$ remains in the air for centuries. [It is nevertheless important to improve efficiency of oil use, because that buys us time to develop technologies and fuels for the post-oil era, and high efficiency surely will be needed in the post-oil era.] However, the point is this: oil will not determine future climate change. Coal will.

Avoiding dangerous atmospheric $CO_2$ levels requires curtailment of $CO_2$ emissions from coal. Atmospheric $CO_2$ can be stabilized by phasing out coal use except where the $CO_2$ is captured and sequestered, as is feasible at power plants. Indeed, agreement to phase out coal use except where the $CO_2$ is captured is 80 percent of the solution to the global warming crisis. Of course, it is a tall order, as coal is now the world's largest source of electrical energy. Over the next few decades those coal plants must be closed or adjusted to capture their $CO_2$ emissions. Yet it is a doable task. Compare that task, for example, with the efforts and sacrifices that went into World War II.

## RESPONSIBILITY FOR GLOBAL WARMING

Responsibility for global warming is proportional to cumulative $CO_2$ emissions, not to current emission rates (http://pubs.giss.nasa.gov/docs/2007/2007_Hansen_etal_1.pdf). This is physical fact, not an ethical statement. It is a consequence of the long lifetime of atmospheric $CO_2$.

## Fossil Fuel CO₂ Emissions

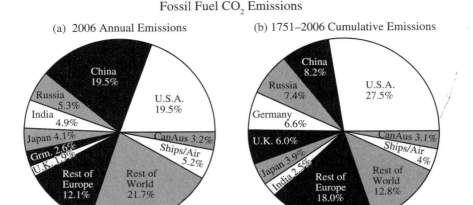

(a)  2006 Annual Emissions                    (b) 1751–2006 Cumulative Emissions

Figure 4. Annual and cumulative fossil fuel $CO_2$ emissions by country of emission (CDIAC data for 1751–2004, BP for 2005–6; cf. Atmos. Chem. Phys. 7, 2287–2312, 2007).

Responsibility of the United States is more than three times larger than that of any other nation (Figure 4). Despite rapid growth of emissions from China, the United States will continue to be the nation most responsible for climate change for at least the next few decades.

It is also useful to examine per capita fossil fuel $CO_2$ emissions. Figure 5a shows per capita emissions for the eight nations with largest total emissions, in order of decreasing total emissions. The United States and Canada have the largest per capita emissions, while emissions of Japan, Germany, and the United Kingdom are half as large per capita.

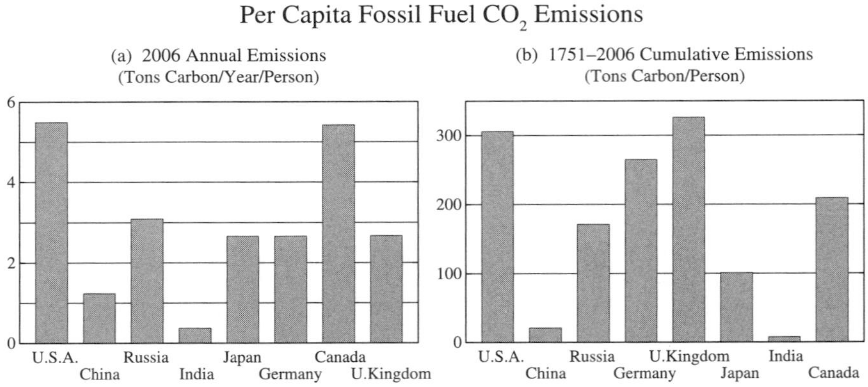

## Per Capita Fossil Fuel CO₂ Emissions

(a)  2006 Annual Emissions          (b)  1751–2006 Cumulative Emissions
(Tons Carbon/Year/Person)                  (Tons Carbon/Person)

Figure 5. Per capita fossil fuel emissions (a) in order of national emissions today, (b) per capita cumulative emission (2006 population) in order of national cumulative emissions (CDIAC data for 1751–2004, BP for 2005–6; cf. Atmos. Chem. Phys. 7, 2287–2312, 2007).

Per capita responsibility for climate change, however, must be based on cumulative national emissions. The United Kingdom has the highest cumulative emissions per capita (2006 population), as shown in Figure 5b. The United States is second in per capita emissions and Germany is third. Increased per capita responsibility of the United Kingdom and Germany is a consequence of their early entries into the industrial era. Recognition of these facts is not an attempt to cast blame. Early emissions of $CO_2$ occurred before the climate problem was recognized and well before it was proven. Yet these facts are worth bearing in mind.

## IMPLICATIONS

Human-made climate change is unambiguously underway. Yet the urgency of the situation is not readily apparent to everybody. Chaotic weather fluctuations mask climate trends, even as climate change alters the nature of weather. Urgency is created by the very inertia of the climate system that delays the effects of gases already added to the air. This delay means that there is additional global warming "in the pipeline" due to human-produced gases already in the air.

Climate system inertia is due in part to the massive oceans, four kilometers deep on average, which are slow to warm in response to increasing greenhouse gases. The effect of this inertia is compounded by positive (amplifying) feedbacks, such as melting of ice and snow, which increases absorption of sunlight, engendering more melting. Such feedbacks are not "runaway" processes, but they make climate sensitive to even moderate climate forcings. [A climate forcing, natural or human-made, is an imposed perturbation of the planet's energy balance. Examples include a change of the sun's brightness or an increase of long-lived greenhouse gases, which trap the Earth's heat radiation.]

Climate inertia and positive feedbacks together create the danger of passing climate "tipping points." A tipping point exists when the climate reaches a point such that no additional forcing is needed to instigate large, relatively rapid climate change and impacts. Impacts of these large climate changes tend to be, overall, detrimental to humans, because civilization is adapted to the relatively stable interglacial period that has existed on our planet for about ten thousand years, and we have settled the land and built great infrastructure within and upon these relatively stable climate zones and coastlines.

Based on current information, we now realize that we have passed or are on the verge of passing several tipping points that pose grave risks for humanity and especially for a large fraction of our fellow species on the planet. This information is gleaned primarily from the Earth's history and ongoing global observations of rapid climate changes, and to a lesser extent from climate models that help us interpret observed changes.

Potential consequences of passing these tipping points include (1) loss of warm season sea ice in the Arctic and thus increased stress on many polar species, possibly leading to extinctions, (2) increasing rates of disintegration of the West Antarctic and Greenland ice sheets, and thus more rapidly rising sea levels in coming decades, (3) expansion of subtropical climates adversely affecting water availability and human livability in regions such as the American West, the Mediterranean, and large areas in Africa and Australia, (4) reduction of alpine snowpack and water run-off that provides fresh water supplies for hundreds of millions of people in many regions around the world, and (5) increased intensity of the extremes of the hydrologic cycle, including more intense droughts and forest fires, on the one hand, but also heavier rains and greater floods, as well as stronger storms driven by latent heat, including tropical storms, tornados, and thunderstorms.

The nearness of these climate tipping points is no cause for despair. On the contrary, the actions that are needed to avert the tipping point problems are not only feasible, they have side benefits that point to a brighter future for life on the planet, with cleaner air and cleaner water. It will be necessary to roll back the airborne amounts of several air pollutants, but that is plausible, given appropriate attention. Already all pollutants except $CO_2$ are falling at or below the lowest IPCC (Intergovernmental Panel on Climate Change) scenarios, and there is much potential for further reductions.

The tendency of the media to continually report bad news on climate change and the human-made factors that drive climate change sometimes paints a picture that is bleaker than that shown by careful analysis. Such information is often misleading about the true status of the Earth, and the impression created may be harmful if it leads to despair about the prospects for achieving a relatively stable climate with a cleaner atmosphere and ocean. I illustrate with data for $CO_2$, the most important climate forcing.

Figure 6 is the "airborne fraction" of fossil fuel $CO_2$ emissions. This is the ratio: the annual increase of $CO_2$ that appears in the Earth's atmosphere (well measured) divided by the annual human emission of fossil fuel

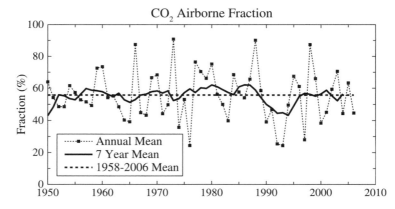

Figure 6. Ratio of observed atmospheric $CO_2$ increase to fossil fuel $CO_2$ emissions (Proc. Natl. Acad. Sci. 101, 16109–16114, 2004).

$CO_2$ into the air (also well known). On average, the increase of $CO_2$ in the air is 57 percent of the fossil fuel emissions. Although this is a large amount, the 43 percent taken up by the ocean, soil, and biosphere is also large. The uptake is large despite the fact that humans are also causing extensive, mostly unwise, deforestation, which adds $CO_2$ to the air. In addition our agricultural practices typically do not encourage storage of carbon in the soil.

There is tremendous potential for reducing atmospheric $CO_2$ via reduction of deforestation, improved forestry practices, and improved agricultural practices that increase carbon storage in the soil. If governments were to encourage such practices, rather than the converse, and if coal use were phased out except where the $CO_2$ is captured, it would be possible to literally roll back the net human-made climate forcing to levels below those defining critical tipping points.

We must remember, at the same time, that the ability of the principal $CO_2$ sink, the ocean, to soak up human-made emissions is limited and slow. If we burn most of the available coal (Figure 3) without $CO_2$ capture, even with the lowest estimates of available coal reserves, it will be impractical if not impossible to avoid passing climate tipping points with disastrous consequences.

## SUMMARY: THE NEED FOR LEADERSHIP

I am optimistic that greenhouse gas emissions can be reduced and atmospheric composition stabilized at a level avoiding disastrous climate effects. My optimism is based in part on the fact that young people are beginning to make their voices heard. They have a powerful effect on our consciences, with an ability to influence policy makers and the captains of industry.

Many individuals are beginning to recognize and appreciate the nature of the climate problem. People want to take actions and they are willing to make sacrifices. However, individual actions cannot solve the problem by themselves.

Based on fossil fuel and carbon cycle facts summarized above, we cannot continue to burn the coal reserves without $CO_2$ capture and sequestration. Solution of this problem can be achieved only via strong government leadership.

Governments must recognize the relative magnitudes of fossil fuel resources, i.e., oil, gas, coal, and unconventional fossil fuels, and they must establish policies that influence consumption in ways consistent with preservation of our climate and life on Earth. The fossil fuel facts dictate essential actions (http://arxiv.org/ftp/arxiv/papers/0706/0706.3720.pdf):

(1) Phase-out of coal use that does not capture $CO_2$. This is 80 percent of the solution, creating a situation in which $CO_2$ emissions are declining sharply. (Coal use will also be affected by the second essential action. Indeed, it is likely that much of the coal will be left in the

ground, as economic incentives spark innovations and positive feed-backs, accelerating progress to the cleaner world beyond fossil fuels.)

(2) A gradually but continually rising price on carbon emissions. This will ensure that, as oil production inevitably declines, humanity does not behave as a desperate addict, seeking every last drop of oil in the most extreme pristine environments and squeezing oil from tar shale, coal, and other high-carbon sources that would ensure destruction of our climate and most species on the planet. Recognition by industry of a continually rising carbon price (and elimination of fossil fuel subsidies) would drive innovations in energy efficiency, renewable energies, and other energy sources that do not produce greenhouse gases.

These are the two fundamental actions that must occur if we are to roll back the net climate forcing and avoid the dangerous climate tipping points, with their foreseeable consequences. Both of these actions are essential.

We can take a long list of supplementary actions that will be needed to avoid hardships and minimize dislocations as we phase into a cleaner world beyond fossil fuels. However, the two essential actions must be given priority and governments must explain the situation to the public.

Supplementary actions include improved efficiency standards on buildings, vehicles, appliances, etc. Rules must be changed so that utilities profit by encouraging efficiency, rather than selling more energy. These changes are necessary for success, and there are many economic opportunities associated with them. Yet governments must realize the essential actions dictated by the physics of the carbon cycle. Specifically, release of $CO_2$ to the air from the large carbon reservoirs, coal and unconventional fossil fuels, must be curtailed.

Further actions will be needed to achieve a rollback of the net climate forcing. These actions (http://arxiv.org/ftp/arxiv/papers/0706/0706.3720.pdf) include reduction of non-$CO_2$ climate forcings and improved agricultural and forestry practices. These actions are important and have multiple benefits, especially in developing countries, but they do not have the great urgency of halting construction of new coal plants without carbon capture. Power plants have long lifetimes, and once their $CO_2$ is released to the air, it is impractical to recover it.

Energy departments, influenced by fossil fuel interests, take it as a God-given fact that we will extract all fossil fuels from the ground and burn them before we move on to other ways of producing usable energy. The public is capable of changing this course dictated by fossil fuel interests, but clear-sighted leadership is needed now if the actions are to be achieved in time.

Tipping points and positive feedbacks exist among people, as well as in the climate system. I believe that the action with the greatest potential to initiate positive feedbacks, and lead to the benefits that will accompany a clean energy future, is a moratorium in the West on new coal-fired power plants unless and until $CO_2$ capture and sequestration technology is available. Such a moratorium would provide the West with sufficient moral authority to sit down with China and other developing countries to find ways, likely including technological assistance, for developing countries to also phase out coal use that does not capture $CO_2$.

Perhaps the most important question is this: can we find the leadership to initiate the tipping point among nations? Can we find a country that will place a moratorium on any new coal-fired power plants unless they capture and store the $CO_2$? Unless this happens soon, there is little hope of avoiding the climate tipping points, with all that implies for life on this planet.

# Bibliography

Abboud, Leila. 2006. "Sun Reigns on Spain's Plains: Madrid Leads a Global Push to Capitalize on New Solar-Power Technologies." *Wall Street Journal*, December 5, A-4.

———. 2007. "How Denmark Paved Way to Energy Independence." *Wall Street Journal*, April 16, A-1, A-13.

Agence France-Presse. 2002. "Norway Says No to Controversial Plan to Store $CO_2$ on Ocean Floor." August 22.

Aksamit, Nicole. 2007. "Young McFoster Had a Farm." *Omaha World-Herald*, August 22, 1-E, 2-E.

Anderson, John Ward. 2007. "Paris Embraces Plan to Become City of Bikes." *Washington Post*, March 24, A-10. http://www.washingtonpost.com/wp-dyn/content/article/2007/03/23/AR2007032301753_pf.html.

Andrews, Edmund L. 2007a. "Bush Makes a Pitch for Amber Waves of Home-grown Fuel." *New York Times*, February 23. http://www.nytimes.com/2007/02/23/washington/23bush.html.

———. 2007b. "Senate Adopts an Energy Bill Raising Mileage for Cars." *New York Times*, June 22. http://www.nytimes.com/2007/06/22/us/22energy.html.

Andrews, Edmund, and Felicity Barringer. 2007. "Bush Seeks Vast, Mandatory Increase in Alternative Fuels and Greater Vehicle Efficiency." *New York Times*, January 24. http://www.nytimes.com/2007/01/24/washington/24energy.html.

Arden, Harvey. 2007. "STAY HOME!!" Email from harvey@harveysplace.net, July 24.

AScribe Newswire. 2004. "Conferences Tackle Key Issues in Air Conditioning, Refrigeration." June 23 (in LEXIS).

Associated Press. 2007a. "Biomass Fuel Plant in Iowa Is Moving Forward," *Omaha World-Herald*, October 5, D-1.

———. 2007b. "Bush Calls for Global Goals on Emissions." *New York Times*, May 31. http://www.nytimes.com/aponline/us/AP-Bush.html.

———. 2007c. "Court Rebukes Administration in Global Warming Case." *New York Times*, April 2. http://www.nytimes.com/aponline/business/AP-Scotus-Greenhouse-Gase.html.

————. 2007d. "Germans Blame Ethanol Boom for—Oh Mein Gott!—Rising Beer Prices." *Omaha World-Herald*, June 3, A-18.

————. 2007e. "Live Earth to Kick Off in Australia." *New York Times*, July 6. http://www.nytimes.com/aponline/world/AP-Music-Live-Earth.html.

————. 2007f. "New York's Mayor Plans Hybrid Taxi Fleet." *New York Times*, May 22. http://www.nytimes.com/aponline/us/AP-Green-Taxis.html.

————. 2007g. "Portugal Celebrates Massive Solar Plant." *New York Times*, March 28. http://www.nytimes.com/aponline/technology/AP-Portugal-Solar-Power-Plant.html.

Auchard, Eric, and Leonard Anderson. 2006. "Google Plans Largest U.S. Solar-Powered Office." Reuters, *Washington Post*, October 16. http://www.washingtonpost.com/wp-dyn/content/article/2006/10/16/AR20061016 01100_pf.html.

Baar, Hein J. W. de, and Michel H. C. Stoll. 1992. "Storage of Carbon Dioxide in the Oceans." In *Arctic Ecosystems in a Changing Climate: An Ecophysiological Perspective*, ed. F. Stuart Chapin III, Robert L. Jefferies, James F. Reynolds, Gaius R. Shaver, and Josef Svoboda, 143–77. San Diego: Academic Press.

Ball, Jeffrey. 2007a. "The Carbon Neutral Vacation." *Wall Street Journal*, July 30, P-1, P-4–5.

————. 2007b. "Climate Change's Cold Economics: Industry Efforts to Fight Global Warming Will Hit Consumers' Pockets." *Wall Street Journal*, February 15, A-12.

————. 2007c. "Green-Fuel Alternative." *Wall Street Journal*, April 16, A-8.

————. 2007d. "The Texas Wind Powers a Big Energy Gamble." *Wall Street Journal*, March 12, A-1, A-11.

Barrett, Joe. 2007. "Ethanol Reaps a Backlash in Small Midwestern Towns." *Wall Street Journal*, March 23, A-1, A-8.

Barringer, Felicity. 2007. "Navajos and Environmentalists Split on Power Plant." *New York Times*, July 27. http://www.nytimes.com/2007/07/27/us/27navajo.html.

Barta, Patrick. 2007a. "Crop Prices Soar, Pushing Up Cost of Food Globally." *Wall Street Journal*, April 9, A-1, A-9.

————. 2007b. "Jatropha Plant Gains Steam in Global Race for Biofuels." *Wall Street Journal*, August 24, A-1, A-12.

Bennhold, Katrin. 2007a. "France Tells U.S. to Sign Climate Pacts or Face Tax." *New York Times*, February 1. http://www.nytimes.com/2007/02/01/world/europe/01climate.html.

————. 2007b. "Paris Journal: A New French Revolution's Creed: Let Them Ride Bikes." *New York Times*, July 16. http://www.nytimes.com/2007/07/16/world/europe/16paris.html.

Betts, Richard A. 2000. "Offset of the Potential Carbon Sink from Boreal Forestation by Decreases in Surface Albedo." *Nature* 408 (November 9): 187–90.

Bjerklie, David. 2007. "Check Your Tires." *Time*, April 9. http://www.time.com/time/printout/0,8816,1603740,00.html.

Bly, Laura. 2007. "How Green Is Your Valet and the Rest?" *USA Today*, July 12, D-1, D-2.

Boudette, Neal E. 2007. "Shifting Gears, General Motors Now Sees Green." *Wall Street Journal*, May 29, A-8.

Bourne, Joel K. 2007. "Growing Fuel: The Wrong Way, the Right Way." *National Geographic* (October): 38-59.

Boyd, Philip W., Andrew J. Watson, Cliff S. Law, Edward R. Abraham, Thomas Trull, Rob Murdoch, Dorothee C. E. Bakker, Andrew R. Bowie, K. O.

Buesseler, Hoe Chang, Matthew Charette, Peter Croot, Ken Downing, Russell Frew, Mark Gall, Mark Hadfield, Julie Hall, Mike Harvey, Greg Jameson, Julie LaRoche, Malcolm Liddicoat, Roger Ling, Maria T. Maldonado, R. Michael McKay, Scott Nodder, Stu Pickmere, Rick Pridmore, Steve Rintoul, Karl Safi, Philip Sutton, Robert Strzepek, Kim Tanneberger, Suzanne Turner2, Anya Waite, and John Zeldis. 2000. "A Mesoscale Phytoplankton Bloom in the Polar Southern Ocean Stimulated by Iron Fertilization." *Nature* 407(October 12): 695–702.

Bradsher, Keith. 2007. "Push to Fix Ozone Layer and Slow Global Warming." *New York Times*, March 15. http://www.nytimes.com/2007/03/15/business/worldbusiness/15warming.html.

Brewer, Peter G., Gernot Friederich, Edward T. Peltzer, and Franklin M. Orr Jr. 1999. "Direct Experiments on the Ocean Disposal of Fossil Fuel $CO_2$." *Science* 284 (May 7): 943–45.

Broad, William J. 2006. "How to Cool a Planet (Maybe)." *New York Times*, June 27. http://www.nytimes.com/2006/06/27/science/earth/27cool.html.

Broder, John M. 2007. "Compromise Measure Aims to Limit Global Warming." *New York Times*, July 11. http://www.nytimes.com/2007/07/11/washington/11climate.htm.

Brown, Lester R. 2006. *Plan B: Rescuing a Planet under Stress and a Civilization in Trouble*, rev. ed. New York: Earth Policy Institute/W. W. Norton.

Browne, Anthony. 2002. "Canute Was Right! Time to Give Up the Coast." *Times* (London), October 11, 8.

Buesseler, Ken O., and Philip W. Boyd. 2003. "Will Ocean Fertilization Work?" *Science* 300 (April 4): 67–68.

Bunkley, Nick. 2007a. "Detroit Finds Agreement on the Need to Be Green." *New York Times*, June 1. http://www.nytimes.com/2007/06/01/business/01auto.html.

———. 2007b. "Seeking a Car That Gets 100 Miles a Gallon." *New York Times*, April 2. http://www.nytimes.com/2007/04/02/business/02xprize.html.

*Business Week*. 2004. "How to Combat Global Warming: In the End, the Only Real Solution May Be New Energy Technologies." August 16, 108.

Butler, Declan. 2007. "Super Savers: Meters to Manage the Future." *Nature* 445 (February 8): 586–88.

Canadell, Josep G., Corinne Le Quéré, Michael R. Raupach, Christopher B. Field, Erik T. Buitenhuis, Philippe Ciais, Thomas J. Conway, Nathan P. Gillett, R. A. Houghton, and Gregg Marland. 2007. "Contributions to Accelerating Atmospheric $CO_2$ Growth from Economic Activity, Carbon Intensity, and Efficiency of Natural Sinks." *Proceedings of the National Academy of Sciences*. Published online before print, October 25.

Carlton, Jim. 2007. "Citicorp Tries Banking on the Natural Kind of Green." *Wall Street Journal*, September 5, B-1, B-8.

Carmichael, Bobby. 2007. "Opposition Takes on Coal Plants." *USA Today*, October 30, 4-B.

Chameides, William, and Michael Oppenheimer. 2007. "Carbon Trading over Taxes." *Science* 315 (March 23): 1670.

Chandler, Michael Alison. 2007. "Without a Car, Suburbanites Tread in Peril: Loudoun Residents Blaze Their Own Risky Trails Where Sidewalks and Bike Paths Are Lacking." *Washington Post*, July 16, B-1. http://www.washingtonpost.com/wp-dyn/content/article/2007/07/15/AR2007071501345_pf.html.

Chisholm, Sallie W. 2000. "Stirring Times in the Southern Ocean." *Nature* 407 (October 12): 685–86.

Clark, Andrew. 2006. "'Open Skies' Air Treaty Threat." *Guardian* (London), February 20. http://www.guardian.co.uk/frontpage/story/0,,1713677,00.html.

Clayton, Chris. 2003. "U.S.D.A. Will Offer Incentives for Conserving Carbon in Soil." *Omaha World-Herald*, June 7, D-1, D-2.

Cline, William R. 1992. *The Economics of Global Warming.* Washington, D.C.: Institute for International Economics.

Clover, Charles. 1999. "Air Travel Is a Threat to Climate." *Daily Telegraph* (London), June 5.

Clover, Charles, and David Millward. 2002. "Future of Cheap Flights in Doubt; Ban New Runways and Raise Fares, Say Pollution Experts." *Daily Telegraph* (London), November 30, 1,4.

Commoner, Barry. 1990. *Making Peace with the Planet.* New York: Pantheon.

Cookson, Craig, and David Firn. 2002. "Breeding Bugs That May Help Save the World; Craig Venter Has Found a Large Project to Follow the Human Genome." *Financial Times* (London), September 28, 11.

Cooper, Glenda. 2007. "Live Earth London's Glacial Pacing: Mixing Music and a Serious Message Gives Concert a Clunky Rhythm." *Washington Post*, July 8, D-4. http://www.washingtonpost.com/wp-dyn/content/article/2007/07/07/AR2007070701201_pf.html.

Cooperman, Alan. 2007a. "Evangelical Body Stays Course on Warming; Conservatives Oppose Stance." *Washington Post*, March 11, A-5. http://www.washingtonpost.com/wp-dyn/content/article/2007/03/10/AR2007031001175_pf.html.

———. 2007b. "Eco-Kosher Movement Aims to Heed Tradition, Conscience." *Washington Post*, July 7, A-1. http://www.washingtonpost.com/wp-dyn/content/article/2007/07/06/AR2007070602092_pf.html.

Cortese, Amy. 2007. "Friend of Nature? Let's See Those Shoes." *New York Times*, March 6. http://www.nytimes.com/2007/03/06/business/businessspecial2/07label-sub.html.

Cowell, Alan. 2007. "Britain Drafts Laws to Slash Carbon Emissions." *New York Times*, March 14. http://www.nytimes.com/2007/03/14/world/europe/14britain.html.

Cox News Service. 2007. "Investors Sizing Up 'Carbon Footprints.'" *Omaha World-Herald*, September 30, D-1.

Coy, Peter. 2002. "The Hydrogen Balm? Author Jeremy Rifkin Sees a Better, Post-petroleum World." *Business Week*, September 30, 83.

Crutzen, Paul J. 2006. "Albedo Enhancement by Stratospheric Sulfur Injections: A Contribution to Resolve a Policy Dilemma?" *Climatic Change* 77, no. 3 (August).

Daviss, Bennett. 2007. "Green Sky Thinking: Could Maverick Technologies Turn Aviation into an Eco-success Story? Yes, but Time Is Running Out." *New Scientist*, February 24, 32–38.

Deutsch, Claudia H. 2007a. "Companies Giving Green an Office." *New York Times*, July 3. http://www.nytimes.com/2007/07/03/business/03sustain.html.

———. 2007b. "Trying to Connect the Dinner Plate to Climate Change." *New York Times*, August 29. http://www.nytimes.com/2007/08/29/business/media/29adco.html.

Dicum, Gregory. 2007. "Plugging into the Sun." *New York Times*, January 4. http://www.nytimes.com/2007/01/04/garden/04solar.html.

Dunne, Nancy. 2002. "Climate Change Research Sparks Hawaii Protests." *Financial Times* (London), June 20, 2.

Easterbrook, Gregg. 2007. "Global Warming: Who Loses—and Who Wins?" *Atlantic*, April, 5–64.

Economists' Letter on Global Warming. 1997. June 23. http://uneco.org/global_warming.html.

Eilperin, Juliet. 2007a. "Beyond Wind and Solar, a New Generation of Clean Energy." *Washington Post*, September 1, A-1. http://www.washingtonpost.com/wp-dyn/content/article/2007/08/31/AR2007083102054_pf.html.

———. 2007b. "150 Global Firms Seek Mandatory Cuts in Greenhouse Gas Emissions." *Washington Post*, November 30, A-3. http://www.washingtonpost.com/wp-dyn/content/article/2007/11/29/AR2007112902039_pf.html.

Environmental Defense. 2007. "More CEOs Call for Climate Action." *Solutions* 38, no. 5 (November): 9. http://www.environmentaldefense.org.

Environment News Service. 2002a. "Global Warming Could Hamper Ocean Sequestration." December 4. http://ens-news.com/ens/dec2002/2002-12-04-09.asp.

———. 2002b. "Hot Polymer Catches Carbon Dioxide." May 29. http://www.ens-newswire.com/ens/may2002/2002-05-29-09.asp#anchor5.

———. 2002c. "Liquid $CO_2$ Dump in Norwegian Sea Called Illegal." July 11. http://ens-news.com/ens/jul2002/2002-07-11-02.asp.

———. 2003a. "Hydrogen Leakage Could Expand Ozone Depletion." June 13. http://ens-news.com/ens/jun2003/2003-06-13-09.asp.

———. 2003b. "Iron Link to $CO_2$ Reductions Weakened." April 10. http://ens-news.com/ens/apr2003/2003-04-10-09.asp#anchor8.

———. 2006a. "British Travel Agents Launch Carbon Offset Scheme." November 28. http://www.ens-newswire.com/ens/nov2006/2006-11-28-05.asp.

———. 2006b. "Italy to Build World's First Hydrogen-Fired Power Plant." December 18. http://www.ens-newswire.com/ens/dec2006/2006-12-18-05.asp.

———. 2006c. "Pearl Jam Offsets Climate Footprint of 2006 World Tour." July 10.

———. 2006d. "Texas Announces $10 Billion Wind Energy Deal." October 3.

———. 2007a. "Australia Screws in Compact Fluorescent Lights Nationwide." February 21. http://www.ens-newswire.com/ens/feb2007/2007-02-21-01.asp.

———. 2007b. "Automakers Join Call for National Greenhouse-Gas Limits." June 27. http://www.ens-newswire.com/ens/jun2007/2007-06-27-09.asp#anchor4.

———. 2007c. "Big City Mayors Strategize to Beat Global Warming." May 15. http://www.ens-newswire.com/ens/may2007/2007-05-15-01.asp.

———. 2007d. "British Columbia to Trim Greenhouse Gases, Go Carbon Neutral." February 14. http://www.ens-newswire.com/ens/feb2007/2007-02-14-02.asp.

———. 2007e. "Building Parks Can Help to Climate Proof Cities." June 12. http://www.ens-newswire.com/ens/jun2007/2007-06-12-04.asp.

———. 2007f. "Bush Orders First Federal Regulation of Greenhouse Gases." May 14. http://www.ens-newswire.com/ens/may2007/2007-05-14-06.asp.

———. 2007g. "Business Leaders Urge Climate Action." February 22. http://www.ens-newswire.com/ens/feb2007/2007-02-22-01.asp.

———. 2007h. "California Air Board Adds Climate Labels to New Cars." June 25. http://www.ens-newswire.com/ens/jun2007/2007-06-25-09.asp#anchor7.

———. 2007i. "China Now Number One in Carbon Emissions; USA Number Two." June 19. http://www.ens-newswire.com/ens/jun2007/2007-06-19-04.asp.

———. 2007j. "Coal-fired Power Plant Blocked in Iowa." October 15. http://www.ens-newswire.com/ens/oct2007/2007-10-15-093.asp.

———. 2007k. "College and University Presidents Pledge Climate Neutral Campuses." February 22. http://www.ens-newswire.com/ens/feb2007/2007-02-22-09.asp#anchor6.

———. 2007l. "EPA Petitioned to Limit Greenhouse Gases from Ships." October 5. http://www.ens-newswire.com/ens/oct2007/2007-10-05-094.asp.

———. 2007m. "Ethanol Production Threatens Plains States with Water Scarcity." September 21. http://www.ens-newswire.com/ens/sep2007/2007-09-21-091.asp.

———. 2007n. "First E.U. Commercial Concentrating Solar Power Tower Opens in Spain." March 30. http://www.ens-newswire.com/ens/mar2007/2007-03-30-02.asp.

———. 2007o. "Fresno Airport Goes Solar in a Big Way." April 13. http://www.ens-newswire.com/ens/apr2007/2007-04-13-09.asp#anchor4.

———. 2007p. "Fueling Jets with Animal Fat." July 18. http://www.ens-newswire.com/ens/jul2007/2007-07-18-09.asp#anchor7.

———. 2007q. "Global Wind Power Generated Record Year in 2006." February 12.

———. 2007r. "Guide Delivers Live Earth Global Warming Survival Skills." June 19. http://www.ens-newswire.com/ens/jun2007/2007-06-19-02.asp.

———. 2007s. "House Approves 25 Percent Renewable Energy by 2025." October 17. http://www.ens-newswire.com/ens/oct2007/2007-10-17-091.asp.

———. 2007t. "Industrialized Countries' Greenhouse Gases Hit Record High." November 20. http://www.ens-newswire.com/ens/nov2007/2007-11-20-02.asp.

———. 2007u. "Indy 500 Race Cars to Run on 100% Ethanol." May 25. http://www.ens-newswire.com/ens/may2007/2007-05-25-09.asp#anchor6.

———. 2007v. "Kansas Gets First U.S. Cellulosic Ethanol Plant." August 28. http://www.ens-newswire.com/ens/aug2007/2007-08-28-097.asp.

———. 2007w. "Mandatory U.S. Greenhouse Gas Cap Wins New Corporate Supporters." May 8. http://www.ens-newswire.com/ens/may2007/2007-05-08-01.asp.

———. 2007x. "Solar Paint and Other Solar Surprises." November 12. http://www.ens-newswire.com/ens/nov2007/2007-11-12-094.asp.

———. 2007y. "States Form Climate Registry to Track Greenhouse Gas Emissions." May 8. http://www.ens-newswire.com/ens/may2007/2007-05-08-09.asp#anchor1.

———. 2007z. "University of California Adopts Green Purchasing, Climate Policies." April 4. http://www.ens-newswire.com/ens/apr2007/2007-04-04-09.asp#anchor5.

———. 2007aa. "U.S. Mayors Seek Federal Help to Protect Climate." November 5. http://www.ens-newswire.com/ens/nov2007/2007-11-05-01.asp.

———. 2007bb. "U.S. Mayors Take the Lead in Fighting Climate Change." June 25. http://www.ens-newswire.com/ens/jun2007/2007-06-25-04.asp.

———. 2007cc. "Volvo First Automaker to Go Carbon Dioxide Free." September 24. http://www.ens-newswire.com/ens/sep2007/2007-09-24-03.asp.

———. 2007dd. "WWF: Stopping Climate Change Is Possible." May 2. http://www.ens-newswire.com/ens/may2007/2007-05-02-01.asp.

Etter, Lauren. 2007a. "In China, Plan to Turn Rice into Carbon Credits." *Wall Street Journal*, October 9, A-1, A-15.

———. 2007b. "With Corn Prices Rising, Pigs Switch to Fatty Snacks." *Wall Street Journal*, May 21, A-1, A-14.

Eunjung Cha, Ariana. 2007. "China Embraces Nuclear Future; Optimism Mixes with Concern as Dozens of Plants Go Up." *Washington Post*, May 29.

http://www.washingtonpost.com/wp-dyn/content/article/2007/05/28/
    AR2007052801051_pf.html.
Everson, Darren, and Anjali Athavaley. 2007. "The Downgrading of Business
    Travel." *Wall Street Journal*, July 3, D-1, D-3.
Fahrenthold, David A. 2007. "'Green' Fuel May Damage the Bay: Ethanol Study
    Has Dire Prediction for the Chesapeake." *Washington Post*, July 17, B-1.
    http://www.washingtonpost.com/wp-dyn/content/article/2007/07/16/
    AR2007071601845_pf.html.
Fahrenthold, David A., and Steven Mufson. 2007. "Cost of Saving the Climate
    Meets Real-World Hurdles." *Washington Post*, August 16, A-1.
    http://www.washingtonpost.com/wp-dyn/content/article/2007/08/15/
    AR2007081502432_pf.html.
Faiola, Anthony, and Robin Shulman. 2007. "Cities Take Lead on Environment
    as Debate Drags at Federal Level." *Washington Post*, June 9, A-1.
    http://www.washingtonpost.com/wp-dyn/content/article/2007/06/08/
    AR2007060802779_pf.html.
Fairfield, Hannah. 2007. "When Carbon Is Currency." *New York Times*, May 6.
    http://www.nytimes.com/2007/05/06/business/yourmoney/06emit2.html.
Fairless, Daemon. 2007. "Biofuel: The Little Shrub That Could—Maybe." *Nature*
    449 (October 11): 652–55.
Fallows, James. 1999. "Turn Left at Cloud 109." *New York Times Sunday Magazine*, November 21, 84–89.
Fargione, Joseph, Jason Hill, David Tilman, Stephen Polasky, and Peter
    Hawthorne. 2008. "Land Clearing and the Biofuel Carbon Debt." *Science*
    319 (February 29): 1235–38.
Fialka, John J. 2007a. "Carbon Curbs Gain Backers." *Wall Street Journal*, February 27, A-8.
———. 2007b. "U.S. Plots New Climate Tactic." *Wall Street Journal*, September
    7, A-8.
Finney, Paul Burnham. 2007. "U.S. Business Travelers Let the Train Take Away
    the Strain." *International Herald-Tribune*, April 24, 16.
Fitzgerald, Michael. 2007. "It Takes Deep Pockets to Fight Global Warming."
    *New York Times*, August 12, Business, 3.
Flannery, Tim. 2005. *The Weather Markers: How Man Is Changing the Climate
    and What It Means for Life on Earth*. New York: Atlantic Monthly Press.
Francis, Justin. 2006. "Should the Responsible Traveller Be Flying?" Responsible-
    travel.com press release, February 10. http://www.responsibletravel.com/
    copy/copy900993.htm.
Frank, Robert. 2007. "Living Large While Being Green." Wealth Report. *Wall
    Street Journal*, August 24, W-2.
Friedl, Randall R. 1999. "Atmospheric Chemistry: Unraveling Aircraft Impacts."
    *Science* 286 (October 1): 57–58.
Gaarder, Nancy. 2007a. "Many Digging Deep for Cheaper Energy." *Omaha
    World-Herald*, May 29, A-1, A-2.
———. 2007b. "Winds of Change Blowing in Iowa." *Omaha World-Herald*, May
    13, D-1, D-2.
Gillis, Justin. 2002. "A New Outlet for Venter's Energy; Genome Maverick to
    Take On Global Warming." *Washington Post*, April 30, E-1.
Goodale, Christine L., and Eric A. Davidson. 2002. "Carbon Cycle: Uncertain
    Sinks in the Shrubs." *Nature* 418 (August 8): 601.
Goodell, Jeff. 2007. "The Ethanol Scam." *Rolling Stone*, August 9, 48–53.

Gordon, Anita, and David Suzuki. 1991. *It's a Matter of Survival*. Cambridge: Harvard University Press.

Gore, Albert, Jr. 1992. *Earth in the Balance: Ecology and the Human Spirit*. Boston: Houghton-Mifflin.

Grant, Paul M. 2003. "Hydrogen Lifts Off—with a Heavy Load: The Dream of Clean, Usable Energy Needs to Reflect Practical Reality." *Nature* 424 (July 10): 129–30.

Griscom-Little, Amanda. 2007. "Detroit Takes Charge." *Outside*, April, 60.

*Guardian* (London). 2003. "Rising Tide: Who Needs Essex Anyway." June 12, 4.

Hakim, Danny. 2004. "Several States Likely to Follow California on Car Emissions," *New York Times*, June 11, C-4.

Hansen, James E. Personal communication, April 12, 2007.

Hansen, James, and Makiko Sato. 2007. "Global Warming: East-West Connections." Unpublished.

Hansen, James E. 2007a. "Coal Trains of Death." James Hansen's E-mail List, July 23.

———. 2007b. "Political Interference with Government Climate Change Science." Testimony before the House Committee on Oversight and Government Reform, 110th Cong., 1st sess., March 19.

Harden, Blaine. 2006. "Tree-Planting Drive Seeks to Bring a New Urban Cool; Lower Energy Costs Touted as Benefit." *Washington Post*, September 4, A-1. http://www.washingtonpost.com/wp-dyn/content/article/2006/09/03/AR2006090300926_pf.html.

———. 2007. "Air, Water Powerful Partners in Northwest; Region's Hydro-Heavy Electric Grid Makes for Wind-Energy Synergy." *Washington Post*, March 21, A-3. http://www.washingtonpost.com/wp-dyn/content/article/2007/03/20/AR2007032001634_pf.html.

Havel, Vaclav. 2007. "Our Moral Footprint." *New York Times*, September 27. http://www.nytimes.com/2007/09/27/opinion/27havel.html.

Herrick, Thaddeus. 2002. "The New Texas Wind Rush: Oil Patch Turns to Turbines, as Ranchers Sell Wind Rights; A New Type of Prospector." *Wall Street Journal*, September 23, B-1, B-3.

Hickman, Martin. 2006. "The Prince of Emissions." *Independent* (London), April 1, 1.

Higgins, Michelle. 2006. "Machu Picchu, without Roughing It." *New York Times*, August 12, Travel, 6.

Hillman, Mayer, Tina Fawcett, and Sudhir Chella Rajan. 2007. *The Suicidal Planet: How to Prevent Global Climate Catastrophe*. New York: St. Martin's/Dunne.

Hitt, Greg. 2007. "Changed Climate on Warming." *Wall Street Journal*, March 20, A-6.

Hoffman, Ian. 2004. "Iron Curtain over Global Warming; Ocean Experiment Suggests Phytoplankton May Cool Climate." *Daily Review* (Hayward, Calif.), April 17 (in LEXIS).

Hord, Bill. 2007a. "Closed-loop Ethanol Plant Plugged." *Omaha World-Herald*, December 1, D-1, D-2.

———. 2007b. "Mead Plant Hailed as 'Revolutionary.'" *Omaha World-Herald*, June 29, D-1.

Houlder, Vanessa. 2002. "Rise Predicted in Aviation Carbon Dioxide Emissions." *Financial Times* (London), December 16, 2.

*Houston Chronicle*. 2007. "Hydrogen Car Has Far to Go." Reprinted in *Omaha World-Herald*, September 9, 2-D.

Hughes, Kathleen. 2007. "To Fight Global Warming, Some Hang a Clothesline." *New York Times*, April 12. http://www.nytimes.com/2007/04/12/garden/12clothesline.html.

Hy-Vee. 2007. "Hy-Vee Seafood Country of Origin." Peony Park Hy-Vee, Omaha, Neb. September 27. Photocopy.

Illinois Environmental Protection Agency. N.d. "Energy and Equity: The Full Report." http://www.globalgreen.org/media/publications/CommDevelopClimate%20ALL.pdf.

*Industrial Environment.* 2002. "E.U. Plans to Become First Hydrogen Economy Superpower." 12, no. 13 (December), n.p. (in LEXIS).

International Energy Agency. 1993. *Cars and Climate Change.* Paris: International Energy Agency.

Jackson, Robert B., Jay L. Banner, Esteban G. Jobbagy, William T. Pockman, and Diana H. Wall. 2002. "Ecosystem Carbon Loss with Woody Plant Invasion of Grasslands." *Nature* 418 (August 8): 623–26.

Johansen, Bruce. 2007. "Scandinavia Gets Serious about Global Warming." *Progressive*, July, 22–24.

Johnson, Keith. 2007. "Alternative Energy Hit by a Windmill Shortage." *Wall Street Journal*, July 9, A-1, A-13.

Johnston, David Cay. 2000. "Some Need Hours to Start Another Day at the Office." *New York Times*, reprinted in *Omaha World-Herald*, February 6, 1-G.

Jordan, Steve. 2007a. "Becoming Greener: Environmental Concerns, Lower Costs Push Drive for Energy-Efficient Buildings." *Omaha World-Herald*, July 22, D-1, D-2.

———. 2007b. "Coal-Emission Cleanup a Challenge for Utilities." *Omaha World-Herald*, October 8, D-1, D-2.

"Kansans Rallied to Resist Coal-Burning Power Plants." 2008. Environment News Service, March 12. http://www.ens-newswire.com/ens/mar2008/2008-03-12-092.asp.

Kanter, James. 2007. "Across the Atlantic, Slowing Breezes." *New York Times*, March 7. http://www.nytimes.com/2007/03/07/business/businessspecial2/07europe.html.

Kaplow, Larry. 2001. "Solar Water Heaters: Israel Sets Standard for Energy; Cutting Dependence: Jerusalem's Alternative Energy Use a Lesson for United States." *Atlanta Journal and Constitution*, August 9, 1-P.

Karen, Mattias. 2006. "Sweden Outdoes Bush with Goal to Go Oil-Free by 2020." Associated Press, *Seattle Times*, February 8. http://seattletimes.nwsource.com/html/nationworld/2002791274_sweden08.html.

Keates, Nancy. 2007. "Building a Better Bike Lane." *Wall Street Journal*, May 4, W-1, W-10.

Kelley, Kate. 2006. "City Approves 'Carbon Tax' in Effort to Reduce Gas Emissions." *New York Times*, November 18. http://www.nytimes.com/2006/11/18/us/18carbon.html.

Kennedy, Donald. 2007. "The Biofuels Conundrum." *Science* 316 (April 27): 515.

Kerr, Richard A. 2006. "Pollute the Planet for Climate's Sake?" *Science* 314 (October 20): 401–3.

Kher, Unmesh. 2007. "Pay for Your Carbon Sins." *Time*, April 9. http://www.time.com/time/printout/0,8816,1603737,00.html.

Kintisch, Eli. 2007a. "Light-Splitting Trick Squeezes More Electricity Out of Sun's Rays." *Science* 317 (August 3): 583–84.

———. 2007b. "Making Dirty Coal Plants Cleaner." *Science* 317 (July 13): 184–86.

———. 2007c. "Scientists Say Continued Warming Warrants Closer Look at Drastic Fixes." *Science* 318 (November 16): 1054–55.

Kleiner, Kurt. 2007. "The Shipping Forecast." *Nature* 449 (September 20): 272–73.

Knoblauch, Jessica A. 2007. "Have It Your (the Sustainable) Way." *EJ* (*Environmental Journalism*), Spring, 28–30, 46.

Kolbert, Elizabeth. 2007. "Don't Drive, He Said." Talk of the Town. *New Yorker*, May 7, 23–24.

Kondratyev, Kirill, Vladimir F. Krapivin, and Costas A. Varotsos. 2004. *Global Carbon Cycle and Climate Change*. Berlin: Springer/Praxis.

Krauss, Clifford. 2007. "Ethanol's Boom Stalling as Glut Depresses Price." *New York Times*, September 30. http://www.nytimes.com/2007/09/30/business/30ethanol.html.

Krugman, Paul. 2007. "The Sum of All Ears." *New York Times*, January 29, A-23.

Kurtz, Howard. 2007. "RM + WSJ: Let's Do the Math." *Washington Post*, May 14, C-1.

Lal, R. 2004. "Soil Carbon Sequestration Impacts on Global Climate Change and Food Security." *Science* 304 (June 11): 1623–27.

Landauer, Robert. 2002. "Big Changes in Our China Suburb." *Sunday Oregonian*, October 20, F-4.

Layton, Lyndsey. 2007. "A Carbon-Neutral House? Plan Would Offset Emissions by End of Current Congress." *Washington Post*, May 25, A-17. http://www.washingtonpost.com/wp-dyn/content/article/2007/05/24/AR2007052402146.html.

Lazaroff, Cat. 2002. "Replacing Grass with Trees May Release Carbon." Environment News Service, August 8. http://ens-news.com/ens/aug2002/2002-08-08-07.asp.

Lazo, Alejandro. 2007. "A Shorter Link between the Farm and Dinner Plate: Some Restaurants, Grocers Prefer Food Grown Locally." *Washington Post*, July 29, A-1. http://www.washingtonpost.com/wp-dyn/content/article/2007/07/28/AR2007072801255.html.

Lean, Geoffrey. 2001. "We Regret to Inform You That the Flight to Malaga Is Destroying the Planet: Air Travel Is Fast Becoming One of the Biggest Causes of Global Warming." *Independent* (London), August 26, 23.

Lester, Benjamin. 2007. "Greening the Meeting." *Science* 318 (October 5): 36–38.

Lippert, John. 2002. "General Motors Chief Weighs Future of Fuel Cells." *Toronto Star*, September 27, E-3.

Lohr, Steve. 2007. "Energy Standards Needed, Report Says." *New York Times*, May 17. http://www.nytimes.com/2007/05/17/business/17energy.html.

Lovelock, James. 2006. *The Revenge of Gaia: Why the Earth Is Fighting Back—and How We Can Still Save Humanity*. London: Allen Lane.

Lovins, Amory. 2005. "More Profit with Less Carbon." *Scientific American*, September, 74, 76–83.

Luhnow, David, and Geraldo Samor. 2006. "As Brazil Fills Up on Ethanol, It Weans Off Energy Imports." *Wall Street Journal*, January 9, A-1, A-8.

Lynas, Mark. 2006. "Fly and Be Damned." *New Statesman* (London), April 3, 12–15. http://www.newstatesman.com/200604030006.

Machalaba, Daniel. 2007. "Crowds Heeds Amtrak's 'All Aboard.'" *Wall Street Journal*, August 23, B-1, B-22.

MacLeod, Calum. 2007. "China Envisions Environmentally Friendly 'Eco-city.'" *USA Today*, February 16, 9-A.

Mallaby, Sebastian. 2007. "Carbon Policy That Works: Avoiding the Pitfalls of Kyoto Cap-and-Trade." *Washington Post*, July 23, A-17. http://www.washingtonpost.com/wp-dyn/content/article/2007/07/22/AR2007072200884_pf.html.

Mankiw, N. Gregory. 2007. "One Answer to Global Warming: A New Tax." *New York Times*, September 16. http://www.nytimes.com/2007/09/16/business/16view.html.

Marris, Emma, and Daemon Fairless. 2007. "Wind Farms' Deadly Reputation Hard to Shift." *Nature* 447 (May 10): 126.

Martin, Andrew. 2007. "In Eco-friendly Factory, Low-Guilt Potato Chips." *New York Times*, November 15, A-1, A-22.

Masson, Gordon. 2003. "Eco-friendly Movement Growing in Music Biz." *Billboard*, March 15, 1.

Masters, Coco. 2007a. "End the Paper Chase." *Time*, March 27. http://www.time.com/time/printout/0,8816,1603633,00.html.

———. 2007b. "Fill'er Up with Passengers." *Time*, March 27. http://www.time.com/time/printout/0,8816,1603736,00.html.

———. 2007c. "Kill the Lights at Quitting Time." *Time*, March 27. http://www.time.com/time/printout/0,8816,1603601,00.html.

———. 2007d. "Make Your Garden Grow." *Time*, March 27. http://www.time.com/time/printout/0,8816,1603649,00.html.

———. 2007e. "Rake in the Fall Colors." *Time*, March 27. http://www.time.com/time/printout/0,8816,1603631,00.html.

———. 2007f. "Shut Off Your Computer." *Time*, March 27. http://www.time.com/time/printout/0,8816,1603535,00.html.

McCrea, Steve. 1996. "Air Travel: Eco-tourism's Hidden Pollution." *San Diego Earth Times*, August. http://www.sdearthtimes.com/et0896/et0896s13.html.

McGuire, Bill. 2005. *Surviving Armageddon: Solutions for a Threatened Planet.* New York: Oxford University Press.

McKibben, Bill. 1989. *The End of Nature.* New York: Random House.

———. 2007. "The Race against Warming." *Washington Post*, September 29, A-19, http://www.washingtonpost.com/wp-dyn/content/article/2007/09/28/AR2007092801400_pf.html.

McNulty, Sheila. 2007. "U.S. Power Generation Answer Is Blowing in the Wind." *Financial Times* (London), April 24, 7.

Michael, Daniel, and Susan Carey. 2006. "Airlines Feel Pressure as Pollution Fight Takes Off." *Wall Street Journal*, December 12, A-6.

Monbiot, George. 2006a. *Heat: How to Stop the Planet from Burning.* Toronto: Doubleday Canada.

———. 2006b. "We Are All Killers: Until We Stop Flying." *Guardian*, February 28. http://www.monbiot.com/archives/2006/02/28/we-are-all-killers.

Morton, Oliver. 2007. "Is This What It Takes to Save the World?" *Nature* 447 (May 10): 132–36.

Mufson, Steven. 2006. "A Sunnier Forecast for Solar Energy; Still Small, Industry Adds Capacity and Jobs to Compete with Utilities." *Washington Post*, November 20, D-1. http://www.washingtonpost.com/wp-dyn/content/article/2006/11/19/AR2006111900688_pf.html.

———. 2007a. "Federal Loans for Coal Plants Clash with Carbon Cuts." *Washington Post*, May 14, A-1. http://www.washingtonpost.com/wp-dyn/content/article/2007/05/13/AR2007051301105.html.

————. 2007b. "Florida's Governor to Limit Emissions." *Washington Post*, July 12, D-1. http://www.washingtonpost.com/wp-dyn/content/article/2007/07/11/AR2007071102139_pf.html.

————. 2007c. "In Microbe, Vast Power for Biofuel Organism's Ability to Turn Plant Fibers to Ethanol Captures Investors' Attention." *Washington Post*, October 18, D-1. http://www.washingtonpost.com/wp-dyn/content/article/2007/10/17/AR2007101702216_pf.html.

————. 2007d. "Nuclear Power Primed for Comeback; Demand, Subsidies Spur U.S. Utilities." *Washington Post*, October 8, A-1.

————. 2007e. "On Capitol Hill, a Warmer Climate for Biofuels." *Washington Post*, June 15, D-1. http://www.washingtonpost.com/wp-dyn/content/article/2007/06/14/AR2007061402089_pf.html.

————. 2007f. "Power Plant Rejected over Carbon Dioxide for First Time." *Washington Post*, October 19, A-1. http://www.washingtonpost.com/wp-dyn/content/article/2007/10/18/AR2007101802452_pf.html.

Mufson, Steven, and Dan Morgan. 2007. "Switching to Biofuels Could Cost Lots of Green." *Washington Post*, June 8, D-1. http://www.washingtonpost.com/wp-dyn/content/article/2007/06/07/AR2007060702176_pf.html.

Naik, Gautam. 2007a. "Arctic Becomes Tourism Hot Spot, but Is It Cool?" *Wall Street Journal*, September 24, A-1, A-12.

————. 2007b. "J. Craig Venter's Next Big Goal: Creating New Life." *Wall Street Journal*, June 29, B-1.

National Academy of Sciences. 1991. *Policy Implications of Greenhouse Warming*. Washington, D.C.: National Academy Press.

"National Ban on New Power Plants Without $CO_2$ Controls Proposed." 2008. Environment News Service, March 12. http://www.ens-newswire.com/ens/mar2008/2008-03-12-091.asp.

New Belgium Brewery. N.d. "Our Story." http://www.newbelgium.com/story.php.

*New Scientist*. 2001. "Planting Northern Forests Would Increase Global Warming." Press release, July 11. http://www.newscientist.com/news/news.jsp?id=ns99991003.

————. 2007a. "Keep Earth Cool with Moon Dust." February 10. http://environment.newscientist.com/channel/earth/mg19325905.300-keep-earth-cool-with-moon-dust.html.

————. 2007b. "Volcanic Approach to Solving Climate Woes Goes Up in Smoke." August 11, 16.

*New York Times*. 2007. "The High Costs of Ethanol." Editorial. September 19. http://www.nytimes.com/2007/09/19/opinion/19wed1.html.

Nordhaus, William D. 1991. "Economic Approaches to Greenhouse Warming." In *Global Warming: Economic Policy Reponses*, ed. Rudiger Dornbusch and James M. Poterba, 33–66. Cambridge, Mass.: MIT Press.

O'Connell, Sanjida. 2002. "Power to the People." *Times* (London), May 20, n.p. (in LEXIS).

Odling-Smee, Lucy. 2007. "Biofuels Bandwagon Hits a Rut." *Nature* 446 (March 29): 483.

*Omaha World-Herald*. 2007. "No Sugar-Beet Answer." Editorial. April 7, 6-B.

Oppenheimer, Michael, and Robert H. Boyle. 1990. *Dead Heat: The Race against the Greenhouse Effect*. New York: Basic Books.

Ouroussoff, Nicolai. 2007. "Why Are They Greener Than We Are?" *New York Times Sunday Magazine*, May 20. http://www.nytimes.com/2007/05/20/magazine/20europe-t.html.

Owen, David. 2007. "The Dark Side: Making War on Light Pollution." *New Yorker*, August 20, 28–33.

Patrick, Aaron O. 2007. "Life in the Faster Lane." *Wall Street Journal*, July 20, W-1, W-4.

Patzek, Tad W. 2006. Letter to the editor. *Science* 312 (June 23): 1747.

Patzek, Tad W., and David Pimentel. 2005. "Thermodynamics of Energy Production from Biomass." *Critical Reviews in Plant Sciences* 24: 327–64. http://petroleum.berkeley.edu/papers/biofuels/uc_scientist_says_ethanol_uses_m.htm.

Pearce, Fred. 2007. *With Speed and Violence: Why Scientists Fear Tipping Points in Climate Change.* Boston: Beacon Press.

Pegg, J. R. 2007a. "U.S. Lawmakers Hear Stern Warnings on Climate Change." Environment News Service, February 14. http://www.ens-newswire.com/ens/feb2007/2007-02-14-10.asp.

———. 2007b. "U.S. Senators Propose Compulsory Greenhouse Gas Cuts." Environment News Service, October 18.

Pollack, Andrew. "Scientists Take New Step Toward Man-Made Life." *New York Times*, January 24, 2008. http://www.nytimes.com/2008/01/24/science/24cnd-genome.html.

Price, Catherine. 2007. "A Chicken on Every Plot, a Coop in Every Backyard." *New York Times*, September 19. http://www.nytimes.com/2007/09/19/dining/19yard.html.

Raymond, Peter A., and Jonathan J. Cole. 2003. "Increase in the Export of Alkalinity from North America's Largest River." *Science* 301 (July 4): 88–91.

Reuters. 1999. "Aircraft Pollution Linked to Global Warming: Himalayan Glaciers Are Melting, with Possibly Disastrous Consequences." *Baltimore Sun*, June 13, 13-A.

Revkin, Andrew C. 2000a. "Antarctic Test Raises Hope on a Global-Warming Gas." *New York Times*, October 12, A-18.

———. 2000b. "Planting New Forests Can't Match Saving Old Ones in Cutting Greenhouse Gases, Study Finds." *New York Times*, September 22, A-23.

———. 2007. "Carbon-Neutral Is Hip, but Is It Green?" *New York Times*, April 29. http://www.nytimes.com/2007/04/29/weekinreview/29revkin.html.

Revkin, Andrew C., and Patrick Healy. 2007. "Coalition to Make Buildings Energy-Efficient." *New York Times*, May 17. http://www.nytimes.com/2007/05/17/us/17climate.html.

Revkin, Andrew C., and Matthew L. Wald. 2007. "Solar Power Captures Imagination, Not Money." *New York Times*, July 16. http://www.nytimes.com/2007/07/16/business/16solar.html.

Richards, Bill. 2008. "A Good Combination: Biofuels, Smart Tilling." *Omaha World-Herald*, March 11, 7-B.

Riding, Alan. 2007. "Stars Join Their Voices to Support Live Earth." *New York Times*, July 8. http://www.nytimes.com/2007/07/08/arts/music/08liveearth.html.

Rifkin, Jeremy. 2002. *The Hydrogen Economy: The Creation of the Worldwide Energy Web and the Redistribution of Power on Earth.* New York: Jeremy P. Tarcher/Putnam.

Roberts, Paul. 2004. *The End of Oil: The Edge of a Perilous New World.* Boston: Houghton-Mifflin.

Rogers, Paul. 2006. "Solar Energy Heats Up." *Omaha World-Herald*, October 15, 1-RE, 2-RE.

————. 2007. "New American Home: A Showcase for Energy Efficiency, 'Green' Design." *Omaha World-Herald*, March 11, RE-1, RE-2.

Rohter, Larry. 2006. "With Big Boost from Sugar Cane, Brazil Is Satisfying Its Fuel Needs." *New York Times*, April 10. http://www.nytimes.com/2006/04/10/world/americas/10brazil.html.

Romm, Joseph J. 2007. *Hell and High Water: The Solution and the Politics—and What We Should Do.* New York: William Morrow.

Rosenthal, Elisabeth. 2007. "Vatican Penance: Forgive Us Our Carbon Output." *New York Times*, September 17. http://www.nytimes.com/2007/09/17/world/europe/17carbon.html.

Rosenthal, Elisabeth. 2008. "Studies Deem Biofuels a Greenhouse Threat." *New York Times*, February 8. http://www.nytimes.com/2008/02/08/science/earth/08wbiofuels.html.

Rowe, Mark. 2000. "When the Music's Over . . . a Forest Will Rise." *Independent* (London), June 25, 5.

Ryall, Julian. 2002. "Tokyo Plans City Coolers to Beat Heat." *Times* (London), August 11, 20.

Samuelson, Robert J. 2007. "Prius Politics." *Washington Post*, July 25, A-15. http://www.washingtonpost.com/wp-dyn/content/article/2007/07/24/AR2007072401855_pf.html.

Sanders, Robert. 2001. "Standby Appliances Suck Up Energy." *Cal Neighbors*, Spring. http://communityrelations.berkeley.edu/CalNeighbors/Spring2001/appliances.html.

Sayre, Caroline. 2007. "Make One Right Turn after Another." *Time*, April 9. http://www.time.com/time/printout/0,8816,1603741,00.html.

Schlermeier, Quirin. 2003. "The Oresmen." *Nature* 421 (January 9): 109–10.

Schiermeier, Quirin. 2008. "Europe to Capture Carbon." *Nature* 451(January 17): 232.

Schlesinger, W. H., and J. Lichter. 2001. "Limited Carbon Storage in Soil and Litter of Experimental Forest Plots under Increased Atmospheric $CO_2$." *Nature* 411 (May 24): 466–69.

Schultz, Martin G., Thomas Diehl, Guy P. Brasseur, and Werner Zittel. 2003. "Air Pollution and Climate-Forcing Impacts of a Global Hydrogen Economy." *Science* 302 (October 24): 624–27.

Schulze, Ernst-Detlef, Christian Wirth, and Martin Heimann. 2000. "Managing Forests after Kyoto." *Science* 289 (September 22): 2058–59.

Schwarzenegger, Arnold, and Jodi Rell. 2007. "States Can't Wait on Global Warming; Lead or Step Aside, EPA." *Washington Post*, May 21, A-13. http://www.washingtonpost.com/wp-dyn/content/article/2007/05/20/AR2007052001059_pf.html.

*Science.* 2007. "Melting Faster" [review of *Geophysical Research Letters* 34, L09501 (2007)]. Editors' Choice. 316 (May 18): 955.

Searchinger, Timothy, Ralph Heimlich, R. A. Houghton, Fengxia Dong, Amani Elobeid, Jacinto Fabiosa, Simla Tokgoz, Dermot Hayes, and Tun-Hsiang Yu. 2008. "Use of U.S. Croplands for Biofuels Increases Greenhouse Gases Through Emissions from Land-Use Change." *Science* 319 (February 29): 1238–40.

Seibel, Brad A., and Patrick J. Walsh. 2001. "Potential Impacts of $CO_2$ Injection on Deep-Sea Biota." *Science* 294 (October 12): 319–20.

Silver, Cheryl Simon, and Ruth S. DeFries. 1990. *One Earth, One Future: Our Changing Global Environment.* Washington, D.C.: National Academy Press.

Sisario, Ben. 2007. "Songs for an Overheated Planet." *New York Times*, July 6. http://www.nytimes.com/2007/07/06/arts/music/06eart.html.

Slater, Dashka. 2007. "Public Corporations Shall Take Us Seriously." *New York Times Sunday Magazine*, August 12, 22–27.

Smith, Rebecca. 2007a. "Inside Messy Reality of Cutting $CO_2$ Output." *Wall Street Journal*, July 12, A-1, A-10.

———. 2007b. "New Plants Fueled by Coal Are Put on Hold." *Wall Street Journal*, July 25, A-1, A-10.

Smith, Rebecca. 2008. "Wind, Solar Power Gain Users." *Wall Street Journal*, January 18, A-6.

Solomon, Deborah. 2007. "Carbon's New Battleground: Lawmakers Favor Cap-and-Trade System, Economists Prefer a Tax." *Wall Street Journal*, September 12, A-4.

Spector, Mike. 2007. "Can U.S. Adopt Europe's Fuel-efficient Cars?" *Wall Street Journal*, June 26, B-1.

Spencer, Jane. 2007. "Big Firms to Press Suppliers on Climate." *Wall Street Journal*, October 9, A-7.

Speth, James Gustave. 2004. *Red Sky at Morning: America and the Crisis of the Global Environment*. New Haven, Conn.: Yale University Press.

Staba, David. 2007. "An Old Steel Mill Retools to Produce Clean Energy." *New York Times*, May 22. http://www.nytimes.com/2007/05/22/nyregion/22wind.html.

Steinman, David. 2007. *Safe Trip to Eden: 10 Steps to Save Planet Earth from Global Warming Meltdown*. New York: Thunder's Mouth Press.

Stoll, John D. 2006. "Visions of the Future: What Will the Car of Tomorrow Look Like? Perhaps Nothing Like the Car of Today." *Wall Street Journal*, April 17, R-8.

Stone, Richard. 2007. "Can Palm Oil Plantations Come Clean?" *Science* 317 (September 14): 1491.

Strassel, Kimberly. 2007. "Ethanol's Bitter Taste." *Wall Street Journal*, May 18, A-16.

Struck, Curtis. 2007. "The Feasibility of Shading the Greenhouse with Dust Clouds at the Stable Lunar Lagrange Points." *Journal of the British Interplanetary Society* 60 (March): 82–89.

Stuber, Nicola, Piers Forster, Gaby Radel, and Keith Shine. 2006. "The Importance of the Diurnal and Annual Cycle of Air Traffic for Contrail Radiative Forcing." *Nature* 441 (June 15): 864–67.

Stukin, Stacie. 2007. "The Lean, Green Kitchen." *Vegetarian Times*, September, 51–54.

Surowieki, James. 2007. "Fuel for Thought." *New Yorker*, July 23, 25.

Thomas, Chris D. 2007. Review of Michael Novacek, *Terra: Our 100 Million Year Old Ecosystem*. *Nature* 450 (November 15): 349.

Tierney, John. 2007. "A Cool $25 Million for a Climate Backup Plan." *New York Times*, February 13. http://www.nytimes.com/2007/02/13/science/earth/13tier.html.

Tromp, Tracey K., Run-Lie Shia, Mark Allen, John M. Eiler, and Y. L. Yung. 2003. "Potential Environmental Impact of a Hydrogen Economy on the Stratosphere." *Science* 300 (June 13): 1740–42.

Tsuda, Atsushi, Shigenobu Takeda, Hiroaki Saito, Jun Nishioka, Yukihiro Nojiri, Isao Kudo, Hiroshi Kiyosawa, et al. 2003. "A Mesoscale Iron Enrichment in the Western Subarctic Pacific Induces a Large Centric Diatom Bloom." *Science* 300 (May 9): 958–61.

Ulrich, Lawrence. 2007. "They're Electric, but Can They Be Fantastic?" *New York Times*, September 23. http://www.nytimes.com/2007/09/23/automobiles/23AUTO.html.

Urquhart, Frank, and Jim Gilchrist. 2002. "Air Travel to Blame as Well." *Scotsman*, October 8, n.p. (in LEXIS).

U.S. Green Building Council. 2008. "LEED Rating Systems." http://www.usgbc.org/DisplayPage.aspx?CMSPageID=222.

Vella, Matt. 2007. "Biofuel or Bust: On the Road with E-85." *Wall Street Journal*, June 19, D-2.

Vinciguerra, Thomas. 2007. "At 90, an Environmentalist from the '70s Still Has Hope." *New York Times*, June 19. http://www.nytimes.com/2007/06/19/science/earth/19conv.html.

Volk, Tyler. 2006. "Real Concerns, False Gods: Invoking a Wrathful Biosphere Won't Help Us Deal with the Problems of Climate Change" [review of James Lovelock, *The Revenge of Gaia*] *Nature* 440 (April 13): 869–70.

Wald, Matthew. 2006. "It's Free, Plentiful and Fickle." *New York Times*, December 28. http://www.nytimes.com/2006/12/28/business/28wind.html.

————. 2007. "What's So Bad about Big?" *New York Times*, March 7. http://www.nytimes.com/2007/03/07/business/businessspecial2/07big.html.

*Wall Street Journal*. 2006. "Global Air Traffic Rose, Easing Fuel Cost Blow." July 2, A-7.

————. 2007. "Carbon Neutral Chic." Editorial. July 9, A-14.

Walsh, Bryan. 2007a. "Get a Carbon Budget." *Time*, April 9. http://www.time.com/time/printout/0,8816,1603651,00.html.

————. 2007b. "Think outside the Packaging." *Time*, April 9. http://www.time.com/time/printout/0,8816,1603638,00.html.

Watson, A. J., D. C. E. Bakker, A. J. Ridgwell, P. W. Boyd, and C. S. Law. 2000. "Effect of Iron Supply on Southern Ocean $CO_2$ Uptake and Implications for Glacial Atmospheric $CO_2$." *Nature* 407 (October 12): 730–33.

White, Joseph R. 2006. "An Ecotopian View of Fuel Economy." *Wall Street Journal*, June 26, D-4.

White, Martha C. 2007. "Enjoy Your Green Stay." *New York Times*, June 26. http://www.nytimes.com/2007/06/26/business/26green.html.

Whitlock, Craig. 2007. "Cloudy Germany a Powerhouse in Solar Energy." *Washington Post*, May 5, A-1. http://www.washingtonpost.com/wp-dyn/content/article/2007/05/04/AR2007050402466_pf.html.

Wigley, T. M. L. 2006. "A Combined Mitigation/Geoengineering Approach to Climate Stabilization." *Science* 314 (October 20): 452–54.

Wilber, Del Quentin. 2007. "U.S. Airlines under Pressure to Fly Greener; Carriers Already Trying to Save Fuel as Europe Proposes Plan." *Washington Post*, July 28, D-1. http://www.washingtonpost.com/wp-dyn/content/article/2007/07/27/AR2007072702256_pf.html.

Will, George F. 2007. "Fuzzy Climate Math." *Washington Post*, April 12, A-27. http://www.washingtonpost.com/wp-dyn/content/article/2007/04/11/AR2007041102109_pf.html.

Williams, Alex. 2007. "Buying into the Green Movement." *New York Times*, July 1. http://www.nytimes.com/2007/07/01/fashion/01green.html.

Williams, Carol J. 2001. "Danes See a Breezy Solution: Denmark Has Become a Leader in Turning Offshore Windmills into Clean, Profitable Sources of Energy as Europe Races to Meet Emissions Goals." *Los Angeles Times*, June 25, A-1.

Williams, Wendy, and Robert Whitcomb. 2007. *Cape Wind: Money, Celebrity, Class, Politics, and the Battle for Our Energy Future on Nantucket Sound.* New York: PublicAffairs.

Woodard, Colin. 2001. "Wind Turbines Sprout from Europe to U.S." *Christian Science Monitor*, March 14, 7.

Yardley, William. 2007. "In Portland, Cultivating a Culture of Two Wheels." *New York Times*, November 5. http://www.nytimes.com/2007/11/05/us/05bike.html.

Zezima, Katie. 2007. "In New Hampshire, Towns Put Climate on the Agenda." *New York Times*, March 19. http://www.nytimes.com/2007/03/19/us/19climate.htm.

# Index

## About the Author

BRUCE E. JOHANSEN is Frederick W. Kayser Research Professor of Communication and Native American Studies at the University of Nebraska at Omaha. He is the author of dozens of books, including *Global Warming in the 21st Century* (Praeger, 2006), *The Global Warming Desk Reference* (Greenwood Press, 2001), *The Dirty Dozen: Toxic Chemicals and the Earth's Future* (Praeger, 2003), *Indigenous Peoples and Environmental Issues* (Greenwood Press, 2003), and *Silenced! Academic Freedom, Scientific Inquiry, and the First Amendment under Siege in America* (Praeger, 2007). Johansen regularly contributes articles on environmental issues to such national periodicals as *The Nation*, *The Progressive*, the *Wall Street Journal*, and *The Atlantic Monthly*.